Conceits

cognition and perception

by
David Cycleback

Horizontal bar that changes tone left to right

Despite appearance, the middle bar does not change in color or tone. If you cover up the image so only the bar is showing you will see this.

Which line is longer?

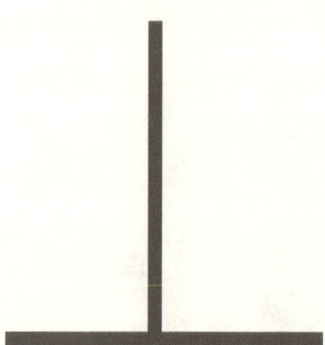

Despite the vertical line appearing longer, the two lines are
the same length. This curiosity is common to most humans.
This illusion happens in the real world, with a telephone pole
or tree appearing taller when vertical than when laying flat
on the ground.

There are societies where perpendicular lines are rare,
such as desert people who live in rounded huts without
perpendicularly angled tables and boxes and television sets.
These people are less likely to be fooled by the above
illusion.   It is interesting that those experienced with
perpendicular lines are those who fall for this illusion. Most
would assume it would be the inexperienced that are more
likely to be tricked.

I remember as a kid walking into the half open bathroom door in the middle of the night. In the dark I saw the moon through the window on the opposite side of the bathroom and assumed the door was wide open. If the door had been closed, hiding the moon, I would have assumed the door was closed and felt for the door knob. This is a case where my assumption was half right: the door was half open. The problem being that the edge of a half opened door hurts your head more than the face of a closed door.

Conceits : Cognition and Perception
by David Rudd Cycleback

Contents

Chapters 1-31 constitute the book Conceits. The end chapters a, b and c are separate short works added later.

==================================

Later Pieces:
   a) Numeration systems and psychology
   b) Movement illusions
   c) Narrative and the perception of still information

Even in Kyoto
how I long for Kyoto
when the cuckoo sings
    -- Basho

Whilst traveling through the Andes Mountains, we
lost our corkscrew and were forced to live on nothing
but food and water for days.
    -- W.C. Fields

1) Conceits

For this book a conceit is defined as a false, artificial, arbitrary, contrived and/or overly simplified rule or set of rules used to explain the way things are or the way they are supposed to be. A conceit is often made to give an answer where the real answer is unknown or to give a simple, convenient answer to a complex situation.

* * * *

A Victorian England book of etiquette stated that on a bookshelf a book by a male author should never be placed next to a book by a female author. The exception was when the authors were married to each other.

* * * *

A wealthy American businessman and amateur historian decided to build a duplicate of an Ancient Greek pillar on his ranch. His expressed intent was to make it as historically accurate as possible, down to the smallest known detail. Partway through the construction scholars discovered that the Ancient Greeks had painted the original pillar a bright light blue. The businessman was taken aback at this finding. All the pillars he had seen in person and depicted in books were unpainted. Painting one of those beautiful stoic pillars a bright color bordered on the distasteful, like following a fine meal with lime jello and cool whip. The businessman built the pillar exact in all known details except it was unpainted.

* * * *

The human being lives in a universe that is mostly beyond its knowledge and comprehension.

None of us knows the volume of the universe, the complete inner workings of our own minds, what birds really think or what it's like to be in someone else's shoes. We can speculate, we can conjecture, we can theorize, but we don't know for certain.

It's fair to assume Albert Einstein would have said there were many areas of science he knew little about. Just because you are a famous nuclear physicist on the cover of *Time* magazine doesn't make you a wiz at biology, veterinary science, economics, geology, forestry and television repair.

* * * *

While humans know little about the universe, they have an innate psychological need for answers and order. Most of us want to know the meaning of the universe and what is our purpose on earth.

* * * *

In an attempt to overcome their lack of knowledge and sate their desires for order, human beings create pseudo answers and artificial order. This is most commonly done with conceits.

* * * *

The following are examples of conceits:

* The sun rises in the morning and sets in the evening. (The sun does not rise and fall. This is a visual illusion caused by humans' position on the surface of a rotating earth.)

* Baby boys should wear blue, baby girls should wear pink.

* When men greet they must shake hands.

* A painting should be framed and hung from the wall. You should not display it on a tabletop or leaned against a wall.

* A Gothic novel must have dark, stormy weather and a castle or mansion.

* It is uncouth to drink wine out of a coffee cup or beer stein. Wine must be drunk from a wine glass.

* A properly set table must have, from left to right, fork on napkin, plate, knife, spoon and drinking glass. A table set another way is set incorrectly.

* A cowboy movie has to take place in a dusty hot place like in Arizona or Texas. If it takes place in Maine, it's not a cowboy movie.

* There is great significance in 10 year (decade) and 100 year (century) increments. Nine, 11 or 98 year durations are of lesser importance.

* * * *

Conceits are used in all facets of our lives. From the fashion rules for the shoes we wear to how we describe the universe to our children. From the way a house is supposed to be decorated to how music is supposed to sound. From the ways we conceptualize the unknown to the required color for artificial turf in a sports stadium. I hate to break it to you sports fans, but there's no practical reason artificial turf can't be blue, purple, grey, red, black or white.

A conceit can be said and unsaid, conscious and nonconscious, innate and learned, known and unknown. In cases it is a set of rules posted on a sign. In other cases it is a gut reaction ('That's just the way it's supposed to be').

Conceits can be trivial ('pencils always go to the right of the pens on my desk') to large (religious, political or philosophical beliefs requiring a leap of faith).

One's conceits can be idiosyncratic or widely held (custom). Many of one's conceits change and develop with time and experiences.

* * * *

Bugs are icky.
For a romantic evening, you need soft music and candlelight.
You must dress up to go to the opera.
Your socks should match in color and pattern.

* * * *

The human is wired to interpret its environment in the form of conceits. The human's environment is so complex, the human constantly bombarded with so much internal and external information, the human uses conceits to create an understandable translation.

Someone who claims to have no conceits has pointed out she has an additional one.

* * * *

Anyone who doesn't believe in the prevalence of conceits should go to a mall or busy downtown street and observe the variety of fashions. And, perhaps more important, observe how he or she reacts to the fashions ('Damn hippie,' 'Must be a Republican,' 'Honey, hide your purse.').

* * * *

For just one day try to live without conceits. No prejudice, no preconceptions, no traditions, no fashion, no habits, no arbitrary choices, no simplified answers to complicated situations, no made up answer when the real answer is unknown, no doing something 'because that's the way I always do it.'

Realize that exchanging one conceit for another is not ridding you of conceits.

If you can't live without conceits for a day, try it for a partial day, try it for an hour, try it for five minutes. Time yourself with your stopwatch.

* * * *

Why is pink so associated with girls and sissies? Is there something inherent about the color, similar to the biological attraction of hummingbirds to brightly colored flowers? Or is it mostly a matter of tradition? If 100 years ago the tradition started that girls wore dark blue, would tough guys today wear pink sweatshirts taunting the guys who wore blue?

* * * *

What is the hair color of your dream lover?

* * * *

If you had to eat maggots and there was no health or taste concern would you rather they were cooked or live? Why?

* * * *

In a dating relationship would you feel uncomfortable if the woman were much taller than the man? Why?

* * * *

Manipulating information

We all purposely limit the amount of information we receive. It's a normal, daily occurrence. The human being doesn't have the mental ability to process everything at once, and must pick and chose what it focuses on.

"Can we discuss this later? I'm busy right now and don't want to lose my concentration."

"Don't anyone tell me the score of last night's game. I had to work and recorded the game so I can watch it tonight."

"Honey, pull the shades. I don't even want to know what the neighbors are doing this time."

"I'm not going to the Doctor, because I don't want to know if there's something wrong with me."

"I'll look at my bank statement on Monday morning. This is the weekend and I want to enjoy myself."

"They're my parents for God's sake. I don't want to hear about their love life."

* * * *

Tricking Yourself

It is probably no surprise to hear that humans trick or otherwise manipulate each other .... Embellishing one's job position to impress the future in-laws .... Psyching out your opponent at the big ping pong tournament .... Tricking your sibling out of the last donut

Humans also trick or otherwise manipulate themselves. Many of the following examples are closely related to the previous *limiting information* examples.

"Honey, hide the bag of Doritos. You know I can't help myself if they're lying around."

"If I buy myself a new power suit, I will have confidence for the meeting."

"I'm going to turn my watch ten minutes ahead so I'm not always so late to meetings."

"I'm going to force myself not to think about her. Maybe that will help heal my broken heart."

Give two examples of how you trick or manipulate yourself.

* * * *

Keeping Up Appearances

We all superficially dress up facts to suit our tastes. Even if we know the meaning remains the same, outer appearances are important.

"I'm not a secretary, I'm an administrative assistant."

"Don't call it a toilet. That's crass. It's a rest room."

"I didn't get a pay raise, better office or the other things I wanted, but I did convince the boss to change my title. You're looking at the new assistant director for data processing. I can't wait to phone mom. She'll be so proud."

"Don't say 'damn.' Say 'darn.'"

What euphemisms do you use?

* * * *

Choosing to pay for what is free

I used to write an email newsletter about collectables. While it had wide readership and received positive feedback, it was nearly impossible to get any donations of time or money to support it. I had planned on having a series of articles on collecting wirephotos-- identification, dating, valuation. Before I was able to include the series, I decided I had enough of doing the newsletter for free and ended it. With the newsletter finished, I computer printed the wirephotos articles into a Spartan 35 page booklet and offered it for sale for about $7 a copy. Within the first week and a half I made more money from that little booklet than I had received in donations in over two years of publishing the newsletter. Because of their bias about how information should be disseminated (physically printed versus email), the readers chose to pay for information they would have received for free. Not that I was complaining.

* * * *

Biases

Most conceits are based on biases. People's views of the world and even of facts are affected by biases.

A bias is a strong preference for or against something for reasons that do not have a rational basis. A bias can be identified when someone is offered the choice of items that are identical except for one subjective quality (color, shape, scent), and the person consistently picks a particular item because of the subjective quality.

> Each morning five shirts are laid out on your bed. The shirts are identical other than in color. If you only or usually pick the blue shirt, you have a bias towards blue, at least as far as the shirts go. If over time you wear all the shirts except the yellow, you have a bias against yellow shirts.

We all have a range of biases. We all have prejudices (meaning, making judgments before all the facts are in, or jumping to conclusions) and predilections (a strong liking or disliking of something based on temperament or prior experience).

While the word bias often has a derogatory connotation, many biases are worthwhile and even helpful. We all have personal preferences that are positive influences on our lives. I feel no need to apologize for preferring Chinese food over Italian, Rachmaninov over Brahms or having a favorite color of blue. No one should run to the confessional because she dislikes watching basketball and loves to wear pearl earrings. Life would be boring without personal preferences.

The problems arise when biases prevent us from being able to make what should be or we represent as rational judgments. Many of our biases make us jump to false conclusions. Many of our biases cloud what should be clear vision. Many people's biases prevent them from seeing the truth right in front of their eyes.

> When there is the latest political scandal, do you in part judge the guilt or innocence based on the accused's political affiliation? Are you more likely to suspect him guilty if he is a member of the other party? If he shares your political beliefs, are you more likely to ascribe the accusations to being partisan attacks?

In the news there are all those latest health findings on what's good for you and what's bad for you: drink this amount of wine weekly, eat this, avoid that, get this amount of exercise. When first hearing the latest health finding, do you in part judge the scientific validity of the report based on how it relates to your lifestyle? If you love red meat are you more likely to accept at face value a report claiming the benefits of red meat and dismiss a report claiming that red meat should removed from one's diet?

When an important medical report is given to the public on television, do you in part judge the validity of the report based on what the doctor is wearing and from where he is presenting the findings? Even if the report is the same, would you give more credence if the doctor is wearing a white lab coat and stethoscope and speaking from a laboratory (test tubes, vials, scientific charts in the background), as opposed to if he is wearing jeans and a well worn T-shirt and speaking from a junky park bench? Why do you think makers of commercials hawking that fad diet or libido pill use actors dressed like doctors in white lab coats?

Many biases are subtle, many are genetic. If we were born cats, we'd have different priorities, different ways of looking at things. We have habits we don't know exist until pointed out by others. Movie makers know that lighting, camera angle and music influences the movie-goers' opinions of the characters.

* * * *

Many people complain that a news organization is biased. Most of these people are not looking for unbiased reporting, but reporting with a different bias (theirs).

* * * *

Killing cockroaches

The traditional way to kill cockroaches is by taking a can of bug spray and spraying the offending creatures. Years back a company invented a different way for killing cockroaches. Instead of directly spraying the bugs, this company had a new disc that was discreetly placed out of sight-- under a bed or refrigerator, the back of a closet. This disc was more effective than the spray can-- meaning, it killed more bugs. The company test marketed the product with inner city single mothers who had cockroach problems in their homes and used bug spray. The mothers were shown how the disc worked and informed it would kill more cockroaches. When polled afterwards, the majority of the women said they would not purchase the disc, as spraying the cockroaches gave them a sense of control.

* * * *

Learning from experience

Former US President Bill Clinton and Vice President Al Gore at a 1997 Press Conference

Much of how the human being sees, interprets and reacts is based on past experience. Both consciously and nonconsciously we use past experience to show us the way things are. Sometimes we learn from repeated experiences, sometimes from a single experience, sometimes from what others tell us.

Most of us have learned not to put our hands on a red burner on the stove because of personal experience (ouch!) or because we were taught. We learn how to identify plants and animals through experience. Many people love to go up and smell roses because they know what roses smelled like before.

Through repetition, or even single experience, many things become second nature. We barely have to be conscious of them. It's raining, reach for the umbrella. If a burner is bright red, don't touch it. Rabbits are soft and sidewalks are hard. Jiggle the handle on the upstairs toilet or the toilet will run all night. Alligators are dangerous. Chocolate is sweet. Salty and fatty foods are bad for you. Grass is green and beaches are sandy.

Our nonconscious minds and bodies learn from experience—depth perception, reacting to gravity, balance. Athletes perfect their skills through practice. By repeating shots and moves, the moves become second nature to the basketball player. Through practice the gymnast gains balance and muscle memory. Juggling and cycling becomes second nature with practice.

Not only do humans learn from precedent, they gain psychological and even physical attachment to what they have learned. This is part of how habits become second nature. If someone was bit by a large dog as a kid she may shake with fear when a large dog approaches her on the street. If someone had childhood vacations at his favorite aunt and uncle's cabin near the beach, he may get a warm feeling when he sees a magazine picture of a similar beach.

This psychological aspect can be helpful. The practical

use of the gut reactions should be apparent in the following: instant fear when a Grizzly crosses your path, uneasy feeling and perhaps even nausea towards a piece of a meat that smells funny and has a strange color, a warm feeling towards someone who gave you a fair shake when no one else would.

The problem is that no matter how seemingly logical or natural or how deep we feel it in our bones, what we learn as correct is not always correct. Often it's dead wrong. Scientists would laugh at laboratory conclusions based on an arbitrary example or hearsay. Yet this is how we learn in everyday life.

Even your eyes can lie. If you don't believe me, <u>take a second look at the earlier picture. It is not of Bill Clinton and Al Gore</u>. Both are Bill Clinton, but one has different hair. Your brain and eyes were in the habit of seeing things a certain way.

Visual illusions illustrate that even our brains have conceits about the way things are. Look at the images on the following four pages.

Spiral

What may appear to be a spiral, is a series of circles. If you carefully trace your finger along a circle, you will see this.

tilted lines

Despite appearance, all the columns are of equal width and parallel to each other.

warped circle

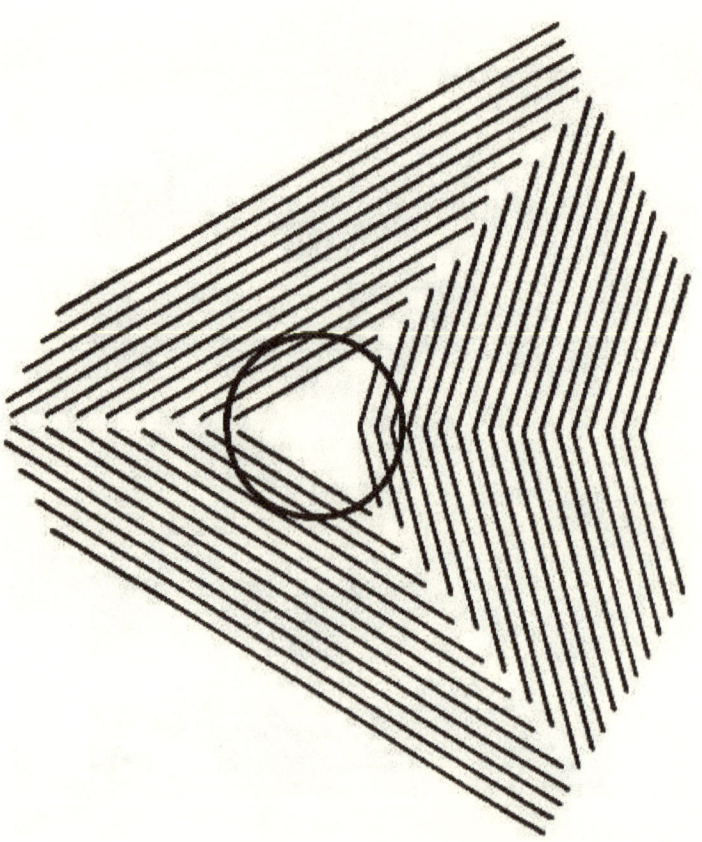

Despite appearance, the above circle is perfectly round. It is the overlapping lines that make it appear warped.

Greta Garbo in hat and coat

There is no illusion with this picture other than caused by your expectation that there was one. It's just a picture of Greta Garbo wearing a hat and coat. You must admit it's interesting that after only several images you created a new (and false) logic. You started a minute ago interpreting as true a false image (Clinton) and ended up with interpreting as false a true image. Fascinating.

* * * *

Absolute statements

Scrutiny reveals the fallacies in our sweeping, absolute statements about society or life or politics or art or sports or television programming.

A liberal mayor may proclaim from the podium, "I am against all forms of racial bias" yet supports racial quotas for school admissions and government contacts. Shouldn't he really say, "I am against all forms of racial bias, except for the areas where I support racial bias"?

A conservative states' rights US Senator may proclaim, "I am for states' rights and against the national government imposing their will on states," then blocks a state from enacting a law he dislikes. Shouldn't the statement more accurately have been, "I am for states' rights and against national government imposing their will on states, except for where I'm not for states' rights and am for national government imposing their will on states"?

Looking closely you will discover that most sweeping absolute statements are not about the person attempting to be factually accurate, but trying to gain power relative to someone or something else. They are rhetorical flourishes. When a brother yells at his kid brother, "You always ruin everything!," he knows the statement is not accurate. However in the middle of a sibling fight the statement "You do many things quite well and mom says you got a B+ on you last French quiz which is quite commendable, but you do mess up a percentage of things on various occasions" doesn't pack the in the heat of the moment punch.

* * * *

When their sports teams clearly are not number one, why do college fans and cheerleaders raise their index fingers and yell "We're number one!"?

Notice this is done in the heat of the moment. During Tuesday morning physics class the student likely won't claim the school's 1-6 basketball team is the best in the nation. However, when you point a television camera on him and his friends during Saturday's game out comes the number one sign.

* * * *

Psychological ties

The human being is an emotional animal ... love, hate, romantic attachment, embarrassment, repulsion, giddiness .... This is part of who we are and how we interpret the world. Emotional interpretation is often more important to humans than facts. Emotions regularly override or temper facts. And who's to say it's always a bad thing. Sticking by family through thick and thin isn't a bad rule of thumb.

For humans it is difficult and often impossible to separate meaning from emotion, facts from emotion, worth from emotion. What is right is supposed to *feel* right. Religious faith involves an emotional attachment to the ideas. There is an emotional connection to the art we love. If there wasn't a psychological reaction to the actors on the screen and their story who would pay good money to sit in the dark theatre for two hours?

No matter how well plotted and witty the dialogue, a movie or novel is deemed unworthy if it doesn't move the critic. "It simply didn't *move* me" or "I didn't connect with the characters" is considered appropriate critical judgment to be a newspaper critic.

Even the most logical of people judge facts by their

aesthetic appearance. An M.I.T. engineering professor will spend hours contemplating what picture and background color should be on his upcoming textbook. He may have a fit if the publisher says the book cover will be hot pink.

A mathematics professor may write and rewrite her equations so they are unsmudged, parallel to the top and bottom of the paper and with attractive margins. Even when the answers are correct, she may reject students' homework that is not similarly neat.

Emotional reactions or states can be good and bad. Most would agree that love for your children, leading you to look out for the best interests, is good. Most would agree that getting a warm feeling from kicking friendly dogs is not good.

Emotional states can alter out landscapes. When we are head over heels in love, a drizzly gray day is gorgeous. When we are unrequited, a rainbow can weigh like lead in the heart.

Mood is an integral part of how we plan our lives ... Getting the mood right for a romantic evening ... Decorating the apartment to make you feel at home after a long day at work.

* * * *

Sanity and Custom

People tend to believe that sanity and insanity are absolute, objective terms, with a medical doctor saying a patient is insane as she would say a patient has a broken arm or skin cancer. The popular and legal definitions of sanity and insanity are based on that society's customs and even fashion. No matter what it is, if enough people are doing it it won't be considered insane behavior.

If you don't believe this, examine what currently socially

acceptable behavior would be deemed bizarre, if not psychotic, if no one else in society did them.

* Decorative body mutilation, such as piercing one's ears and getting a tattoo

* Lying in the sun with the expressed intention of turning brown

* Taxidermy

* Wearing makeup and styling and coloring one's hair

* Taking an animal as a pet, giving it a name, walking it around the neighborhood on a leash and telling people it's the new member of the family

* Expecting people to shake your outstretched hand when you meet, and acting slighted by those who don't

* Manicuring one's lawn and garden, including cutting the shrubs into shapes

If you did all of these, and they were not done by anyone else, you would be considered mentally ill and in need of serious medical help.

* * * *

Given once a year to a single college football player, the Heisman Trophy is the most famous sports trophy in the United States. Unknown to each other, two former Heisman Trophy winners and their families lived in the same neighborhood. One afternoon, one of the men's sons came home disappointed in his dad. His dad had always told him how rare was the Heisman Trophy on the living room mantle, but the dad of the kid down the street had the same trophy.

* * * *

New environments

The BaMbuti Pygmies of Congo traditionally live their entire lives in the dense rainforest, where the furthest away anyone can see is feet. They learned, loved, played and hunted in this environment. In his 1961 book *The Forrest People (Touchstone)*, anthropologist Colin Turnbull wrote how he took one of these Pygmies, named Kenge, for his first time to a wide open plain. As the two stood on a hill overlooking the flat land, a group of water buffalo was seen a few miles away. Having no experience of how things appear smaller over long distance, Kenge asked what kind of insects they were. Turnbull told him they were buffalo and Kenge laughed loudly at the "stupid story." Turnbull drove Kenge towards the buffalo. Watching the animals growing visually larger, Kenge became scared and said it was witchcraft.

Human beings develop an idiosyncratic logic and sensibility distinct to the environment where they were brought up. The environment one grows up in is seemingly the world. A kid born and raised in the inner city versus the country, rich versus poor, in Cairo versus Chicago, conservative family versus liberal, woods versus desert. The person who has lived her whole life in Portland or Cairo may get a chuckle at that story about the Pygmy then dismiss the idea that a similar incongruity could exist with her native logic.

As Kenge interpreted the open expanse based on his jungle experience, humans interpret such esoteric and largely unfathomable things as the afterlife and the meaning of the universe based on their limited experience. It should not be surprising that common human interpretation of the supernatural largely has an earthly sensibility. The supernatural beings often dress like humans, live in night and day, drink and eat human-style meals, speak and read and write, play human-style instruments and games, and even sneeze. It should not be a surprise that to the Ancient Egyptians the gods dressed like Egyptians and to the Ancient Greeks the gods dressed like Greeks.

* * * *

Which black rectangle is larger?

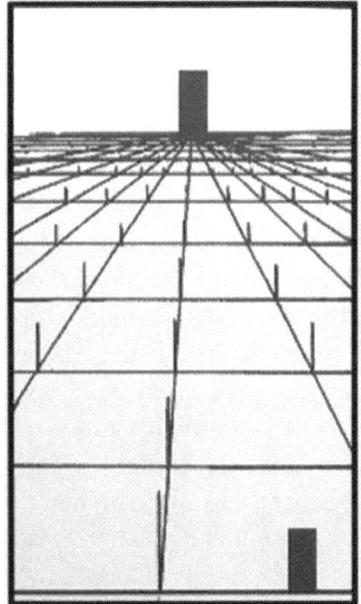

The two rectangles are the same size. Measure them yourself. It is your lifelong experience with diminishing scales in open spaces that caused you to perceive the upper rectangle as larger. Kenge would not have been fooled by this illusion.

* * * *

John Nash's Aliens

John Nash is a famed mathematician and winner of the 1994 Nobel Prize for Economics who was paranoid schizophrenic. While an instructor at M.I.T. and Princeton, Nash suffered severe mental episodes and dropped out of society. He began hallucinating, hearing voices in his head. In this state he deduced that aliens were talking to him.

To most of us his conclusion seems loopy. It does reveal how the human mind works. In a situation well beyond his experience he wanted a concrete answer for what was happening. While bizarre, the aliens conclusion 'logically' matched his illogical situation. It is abnormal to hear voices in one's head so, when one starts hearing voices, normal everyday answers will not explain. It can be expected that someone will explain the abnormal with an abnormal answer, especially when he is in a confused mental state. It is likely no coincidence that Nash nonconsciously picked a conceit that was part of popular culture.

Years later when he had largely recovered from his mental problems, Nash was asked how he had come to the conclusion that aliens were talking to him. He said that he came to conclusion in the same intuitive way that he came to the mathematical conclusions that won him the Nobel Prize.

2) Useful Conceits

While conceits have inherent limitations and pitfalls, many have practical uses and we couldn't function without them. The following are just a few examples.

* While having a law that requires automobiles to drive on the right instead of the left side of the road involves arbitrariness, the usefulness of the law should be clear. This is a case where the powers that be had to pick one side or the other, perhaps flipping a coin. In Britain the coin must have fallen differently.

* Before I go to bed, I make sure my keys are placed in the middle of the kitchen table. I came up with this rule after several hectic mornings looking for the damn things.

* Standard games and sports require conceits to work. Whether an American football field is supposed to be 95, 100 or 105 yards long, you have to pick a length so it is known when someone has scored a touchdown ... Whether a basketball field goal layup counts as 1, 2, 4, 5 or other points, you have to pick an amount before the game starts ... For a tennis match and soccer game, it has to be agreed upon whether *on-the-line* is in bounds or out of bounds.

* One of the most important uses of conceits is they can save us time. Your pre-set rules for clothes you put on in the morning involve personal biases and arbitrary choices. However, without these and other trite rules you might be

unable to leave your house before 4 p.m. This would be a problem if your job starts and 9 a.m. and you're supposed to pick of your kids from kindergarten at 3 p.m.

* * * *

Many conceits don't have practical uses, but are harmless feel goods. Sometimes feeling good is practical, such as when your doctor has suggested you lower your blood pressure.

   * If you grew up dreaming your house would have a white picket fence and a big oak tree, there's nothing wrong with putting up a white picket fence and planting an oak tree on the land you bought.

   * If you just bought a sports car and think that it should have bold racing stripes, there's nothing wrong with asking the dealer to add bold racing stripes.

   * If you need the theatre effect to enjoy a movie and set up your entertainment room in the basement to look like a movie theatre with theatre seats and a popcorn machine, that sounds cool to me.

If a human perceives a person in a magazine picture and a dog does not, which animal is demonstrating better visual perception? Humans sometimes use as evidence of a dog's dimwittedness that the dog 'doesn't see' the human being in a magazine or book, when, of course, there isn't a human being on the page. It's paper and ink. From its sense of smell alone, the dog knows there ain't no human there. The dog is faulted for not seeing what isn't there.

3 : Human Achievement

Humans use conceits, biases and imaginary environments to reach higher levels of achievement. This achievement can range from a musician composing a great symphony to a ten year old improving her math scores.

Humans do not have the capacity to effectively focus on a variety of tasks simultaneously. To reach higher levels of achievement in an area, the human must put most to all of its focus on that area. Humans must eliminate or stabilize (make a non factor) areas that distract from the needed focus.

This is comparable to a water kettle with four equal sized holes in the top. When water is boiled inside, steam will raise a height from the holes. If three of the holes are sealed, the steam will rise much higher from the remaining hole.

* * * *

The following are everyday examples of manipulating one's mental and physical environment to produce achievement:

* While background music or others' chitchat may be fine while browsing a glossy magazine, many of us cover our ears in order to comprehend a difficult passage or perform a math problem.

* To expand one's mind by meditation someone focuses on a repeated mundane and often arbitrary task, such as following one's breath or repeating a word.

* To improve the team's horrid free throw percentage, the junior high basketball coach teaches the players to focus on the basket and their shooting motion and to ignore the crowd. He has them practice by ignoring recorded crowd noise and cardboard cutouts of fans.

* Many with a fear of speaking reduce their nervousness by imagining the audience wearing only their underwear. They create a fantasy.

* * * *

The Rituals of Baseball

Many consider hitting a baseball to be the most difficult feat in sport. The batter swings a stick to hit a small ball. The thrown ball can reach speeds of over 100 miles per hour. Early 1900s player Ty Cobb holds the record for the highest career batting average in Major League Baseball history. His batting average was 0.367, or 3.67 hits per every 10 turns at bat. Even the greatest hitters fail more than they succeeded. Enough to give anyone a complex.

Ty Cobb at bat in 1908

Baseball hitters, and baseball players in general, are notorious for their strange conceits. Players often wear the same unwashed undershirt and socks during a hitting streak. Most players don't step on the white foul lines when entering and leaving the field. Pitcher Turk Wendell waved to left field every time he entered and left a game. When coming to bat, Nomar Garciaparra goes through a well documented ritual of pulling at his shirt, opening and closing the Velcro straps on his batting gloves and tapping the toes of his shoes. Lucky charms, bracelets, necklaces, gum brands abound the game. Five time batting champion Wade Boggs ate chicken before every game. U.L. Washington batted with a toothpick in his mouth. After parents complained that kids might emulate the unsafe habit, he switched to a q-tip. After the first slump, U.L. was back to the toothpick.

Though many of the rituals are comical, they can aid performance. Hitting requires a calm and focused mind and exceptional mind body coordination, all while the player is surrounded by television cameras, screaming fans and the other pressures of being a professional athlete expected to perform. If wearing the lucky undershirt or repeating an odd ritual eases the batter's mind and gives confidence, it can increase the player's batting average. U.L.'s reason for switching back to a toothpick was because it made him feel more comfortable. While a toothpick as aid may seem nonsensical, the desire to be comfortable makes sense.

* * * *

Faith
For a conceit to aid performance, the person must have faith in the conceit.

During a meditation session, one must accept that the thing of mental focus is worthy (breath, mantra, stone, other). Whether the thing was carefully chosen by an instructor or picked in a rush (a pebble hastily grabbed from the ground), meditation requires you to focus on that thing. If you fret about whether or not the mantra was the perfect pick, this very fretting makes the meditation session less effective.

The lucky blue undershirt only helps the baseball player if he believes it lucky. If the blue undershirt is deemed lucky because he had a great game the first time he wore it, this illustrates the arbitrariness in conceits. If before that big game he pulled his grey undershirt from the drawer, it likely would be the grey undershirt that is considered lucky.

* * * *

Positive achievement is regularly based on false beliefs
There are regular cases where positive achievement is achieved from a false belief. This includes in your daily life. Believing the false, if only temporarily, is a technique we all use to remove distracting thoughts. The following are two examples.

* A placebo helps when the patient falsely believes it is medicine. When the patient knows what it is, a placebo doesn't help.

* A freshman at the University of Georgia, Jessica is entering final exam week before winter break. Unknown to her, her beloved 14 year old cat Tiger just died back home in Savannah. The night before her first test she has her weekly telephone conversation with her parents back home. Jessica asks how Tiger is doing. Her mother says Tiger is doing just fine, adding that the cat is playing with a toy on the couch. After hanging up, Jessica's

mother feels bad about lying, but thinks it was best considering the exams. After a productive week, Jessica takes a bus home to Savannah where her parents break the bad news and explain why they delayed it. Jessica understands, agreeing that the news would have distracted her from her studies.

In both these cases it was a false belief that lead to the desired achievement. In both cases, knowledge of the truth would have hindered the achievement.

This shows that positive achievement arising from a belief is not proof that the belief is correct.

Patients who get better after taking a placebo often swear the pills had to be medicine. To them, getting better is the proof. Even when the doctor informs them it was a placebo, some patients continue to believe it was medicine because they got better.

A sincere faith involves a psychological, often irrational. attachment to the ideas. This psychological aspect is both what helps the placebo-taking patient get better (Most doctors believe positive 'I am getting better' thinking aids recovery) and what prevents him from accepting his belief as false even when confronted with the facts. This psychological attachment has both a positive and a negative result.

* * * *

This points to the fascinating relationship humans have with facts. A human cannot function as it desires without the distortion and suppression of facts.

Even a search for the truth requires false beliefs to focus mental attention. In other words, a search for the truth requires lies.

* * * *

Olympic psychology

For world class Olympic athletes a common rule is that one must believe one is going to win in order to win. Paraphrasing a top speed skater interviewed the day before an Olympic race, "You shouldn't just *think* you will win, you must *know* you will win." In a track, swim or bike race, the difference between first and fourth may be a fraction of a second, and the winning psychology can mean the difference between a win and loss. Of course most of these athletes who are sure they will win will not win, and those who win do not win every time. Even when the belief turns out to be wrong, it may better the athlete from, say, fifth to third or third to second.

* * * *

Whether the idolized is a sports coach, historical leader or artist, most worshipers of a human being worship an unreal representation. Much of the misrepresentation is intentional, followers embellishing good qualities and glossing over bad.

At first it seems strange that groups intentionally misrepresent the person they supposedly idolize. However, similar to sweeping absolute statements mentioned in the first chapter, the representations aren't about complete factual accuracy. Amongst other things, they are concerned with gaining and maintaining members' loyalty and spirit, group self importance and gaining power versus other groups. The word *idolizes* implies the act of changing, changing something into an idol.

It should not surprise that during a political election supporters put their candidate in the best light and their competitor in the worst. Their representation isn't about truth, it's about winning the election. If you ask either campaign manager why he doesn't include bad facts about his candidate in the campaign literature, he'll look at you as if you are crazy.

The psychology of expectations

If your pick up an apple, you expect it to have the taste and consistency of an apple. Even if you love banana cream pie, if the first apple bite has the taste and consistency of banana cream pie you likely will be repulsed and spit it out.

(4)
Limitations of Art

An art form (novel, rock song, painting, poem, movie) is a form of language. The artist uses an art form as a means to communicate an idea or ideas to the audience. I use the word idea in a broad sense, ranging from factual idea to emotional state. I use the term audience to mean whoever is watching the movie, reading the book, listening to the music or viewing the painting. An audience can be one million and it can be one.

To be art a work must be profoundly beautiful or sublime to the audience. It must give an audience a sublime or profoundly beautiful experience. Beauty and sublime cannot be translated into simple words, so I do not define them here.

* * * *

Not only is an art form made up and surrounded by a maze
of conceits, but each form is itself a conceit. This means that
art has both the practical benefits and the inherent limitations
of all conceits.

An artistic conceit can be deep, trivial, traditional,
ephemeral, regional, worldwide, conflicting and so on.

* * * *

The follow are examples of artistic conceits. Notice that
some aren't about the art itself but how the art is presented.

* The way a country music song is supposed to sound.
  What instruments are supposed to be used and what
  instruments should not be used. How the musicians
  should dress and move in a music video. What topics the
  lyrics should cover. What topics the lyrics should not
  cover.

* Don't tell me that you or others don't judge a book by its
  cover. If the cover for a tough guy American football
  star's autobiography was changed from dark blue to
  pink, it would affect sales even though the text remained
  the same.

* Say the Chicago Symphony comes to town and offers
  wonderful performances of Beethoven's 9th Symphony
  and Hayden's Water Music. Many in the audience,
  including perhaps the local newspaper critic, will be
  unable to get beyond the fact that the orchestra dressed
  overly casual. The director in tank top and cutoff jeans.
  The lead violinist in bathrobe and stocking feet. Some in
  the audience will demand their money back, the
  newspaper critic might spend half her review
  complaining about the musicians' clothes.

* The clichéd structure, chords, riffs, chorus-to-lead, ending and starting styles, duration and other conceits of rock 'n roll songs. Upon analysis, you will find that singles by Pat Boone, Black Flag, ABBA, Black Sabbath and John Denver have far more in common than many of the respective fans would be willing to admit.

* A movie must be about people or things that are people-like. A movie about a birch tree would be a bomb. However, you might sell some tickets if you have an animated birch that can walk and talk, wears pants and a shirt, has a good sense of humor and has romantic feelings for that spruce of the opposite sex. If you stick in a car chase or two, an evil woodsman and his bad tempered chainsaw who wants to turn Mr. Birchy and all his tree pals into kindling, a fitting musical score and a happy ending with the woodsman foiled and Mr. Birchy and Miss Spruce smooching under a rainbow with nearby supercute bunnies giggling, you might have a blockbuster on your hands.

* When you go to an art museum, what should it look like inside? What should it not look like? What would be your reaction be if a show of Rembrandts had the original, centuries old paintings displayed in funky neon green and day glow yellow frames?

* In Western culture what art forms are generally considered more artistically significant than others? Novel versus comic book, oil painting versus finger painting, television show versus in theatre movie, classical music versus rock 'n roll, drama versus comedy, violin versus banjo? Why?

* * * *

In order to effectively communicate the essential artistic meaning, the artist must follow most of the audience's conceits. This not only includes the deeper conceits but the shallow.

To have the audience focus on the intended meaning, the artist must be faithful to, or at least take into consideration, most of the audience's expectations. Breaking a convention is a shock, a distraction. If the artist breaks all the conventions the audience will be too distracted to focus on the meaning. If you turn a busy street corner and a nude man painted orange and walking a deer in a tuxedo asks you for directions to the library I bet you won't comprehend the first sentence or two that comes out his mouth no matter how clearly he speaks. Similarly, if you display a Rembrandt painting in a hot pink and lime green fuzzy frame with flashing neon lights and dangling felt dice, don't be surprised if the gallery patron is unable to focus on the painting. If you want the patron to focus on the painting, you use a frame that fits his or her expectations.

* * * *

Artists intentionally bend or break some conventions while following the others. They follow all the other conventions in order to focus the audience's attentions on the intentionally bent or broken convention. I dare you to find a popular *shock rock* band that, while having a disturbing twist, does not follow the majority of fashionable conceits, even those used by The Kingston Trio and Sonny and Cher. What you intend to be shocking can't be shocking, or its shock value will be diluted to water, if the audience's attention is distracted by other things. Totally bewildering is rarely as haunting as a perverse twist of the ordinary.

The juxtaposition of the unexpected with the expected,

the abnormal with the normal, is a common artistic technique. Many movies spend the first portion of the work merely setting up an artificial plot and setting to later subvert. How many monster movies start as a normal everyday white picket story? How many thrillers start as an everyday guy going about his everyday business?

The *theme and variation* is a standard musical technique-- altering the melody the second and third time around in a song or other work of music. In comparison to the remembered theme, the altered variation produces a psychological, sometimes poignant effect for the listener. Music can be plotted in a surprisingly similar way to a movie or novel.

* * * *

No matter how shallow the conceits, the successful artist must use or at least address most of the conceits of the audience. Successful art is a compromise between the artist and the audience. It is a communication and communication requires a common language. The artist may have radical things to say, but must communicate in a form the audience can understand. No matter how profound the meaning, the novelist who ignores all the audience's expectations and sensibilities might as well write the book in a foreign language. Great artists are often keenly aware that much of their artistic vision can never be communicated to others.

> And this is how I sometimes think of myself, as a great explorer who has discovered some extraordinary land from which he can never return to give his knowledge to the world.
>
> -- Geoffrey Firmin in Malcolm Lowry's *Under the Volcano*

Locating Space

Humans often perceive cosmic space as being above their heads, pointing up when asked where it is. Of course space is all around us, including straight below our feet. Even if just a matter of custom, this pointing up is traceable to an ancient human mindset that defines existence by visibility; meaning, something exists where it can be seen and doesn't exist where it can't.

Note the common phrase, "You have to see it to believe it."

(5)
The Impossibleness in Translating Poetry

Beyond the changed words, the foreign language translation of a poem alters and often destroys the original poem. With rare exception the translation of a beautiful poem can be similarly beautiful or literally faithful, but not both.

Poetry is uniquely tied to the native language-- the unique word definition, culture, diction, rhyme, sound, meter, feel and even physical length of words and phrases. Due to the literal and figurative differences between languages, a foreign language translation of a poem not only changes the literal words but the poem. It is not possible to change the language and perfectly preserve the original meaning.

This is elementally illustrated by the translation of simple rhyming poems. While 'dog' and 'fog' rhyme, the standard Spanish translations of 'perro' and 'neblina' do not. To make the translation rhyme, the translator must take liberties with the literal meaning. To keep intact the literal meaning, he must omit the rhyming.

In order to preserve artistic meaning, many translators consciously dismiss literal translation. The translation is often as much the artistic creation of the translator as it is of the original poet.

The reader of a translation is not reading the original poem. The translation may be closely related and beautiful and profound, but it's something different. This illustrates the problem with those who take literally modern translations of ancient texts.

(6)
Presenting works of art 'authentically'

Similar to the problem with translating poetry is the problem in trying to present old works of arts in modern times.

Many wish to present a Shakespeare play or Verdi Opera the way it was originally presented, and there are complaints about colorizing old black and white movies.

Advocates of original presentation often refer to a work of art presented in the original manner as being "authentic."

There are a variety of problems in the presentation of old works. For example, the original work or presentation can be unrealistic to its subject. Shakespeare's plays were written for and originally performed by male actors only. Juliet and Ophelia were performed by boys dressed as women. Even those who like the idea of original presentation prefer the inclusion of actresses, meaning they want a Shakespeare performance modernized.

A similar case is where a grandfather clock chimes in Shakespeare's *Julius Caesar*, yet the grandfather clock had not yet been invented in Caesar's time. Some would argue that fixing this historical error would make the play more historically authentic. Others would counter that, while the grandfather clock clearly is a historical blooper, the play was intended as a work of art not a historical document, and 'fixing' every detail could lessen the play artistically. They might point out that a Paul Cezanne painting of an apple is supposed to represent an apple not look like an apple photographed, and those who criticize the painting for not being photorealistic miss the point.

Technical modernization can improve the audience's perception of an old work. Improved technology makes *Gone With The Wind* look and sound clearer in the theatre today than in 1939. It would be a safe bet that Paul McCartney prefers listening to The Beatles on a CD player rather than on a 1965 record player. Listening to the 1965 record player is more authentic to a fan listening to the music on 1965 record player, but listening to a CD is more authentic to the music itself.

I'll bet you that some old time Beatles fan has an unplugged vintage record player sitting on top of a CD player. This way he gets the old time look and the modern sound.

Presenting an old work must take into context the audience, its culture and sensibilities. A play, movie, novel or painting is continually presented to a modern audience. The language of Shakespeare was the language of the original audience. It is not the natural language of today's audience. Today's audience experiences the play differently. The use of boy actors in female parts won't be viewed in the same way as an original audience viewed it. Boys playing girls and women would at the least distract most to all in a modern audience.

Even when presented 'authentically' (as originally presented), the modern audience won't perceive an old work of art authentically, as they won't experience it as the original ('authentic') audience did. Ironically, making modernizations can make the modern audience's experience closer to the original audience's experience. Making a work newer on one level can make it older on another.

Some recreations are less concerned with the art than the history. Even if the sound is considered unorthodox to modern ears, performing a Mozart symphony using period instruments, hall, dress and manners can be of enlightenment and enjoyment to a modern audience, especially if the audience itself participates in the recreation by dressing and acting historically.

(7)
Visual Illusions: Introduction

'Water in the road' mirage

Visual illusions have tricked and fascinated humans for thousands of years. They have influenced history, religion and society, been studied by scientists and philosophers, used by athletes, architects, medical doctors, engineers and artists, and have amused kids and mystery lovers of all ages. Visual illusions include rare spectacular atmospheric events and mundane everyday moments.

Learning about visual illusions and how they work show us that reality and human perception of reality are different things.

What are Visual Illusions?

A visual illusion is when the viewer misperceives what she is looking at. What she thinks she sees is different than what she is looking at.

Visual illusions include misidentifying objects, perceiving things that don't exist or not perceiving things that do exist. It also includes significantly misjudging qualities, such as color, angle, amount, shape, weight, size and distance. Visual illusions happen to birds, fish, flies, dogs and other animals.

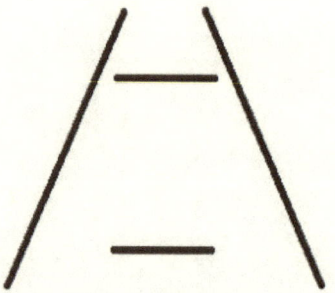

The above horizontal lines are the same length.

Humans have natural and learned ways of perceiving. These methods are good, serving our practical day-to-day needs, but are far from perfect. Mistakes, often minor, are made daily by all humans.

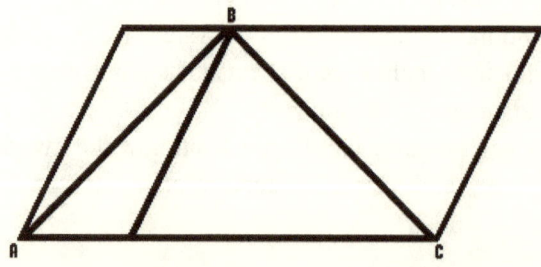

lines AB and BC are the same length

* * * *

Visual illusions are caused by a wide variety of factors. The factors differ from illusion to illusion, and there are multiple causes working together for each illusion. General factors include:

Physiology: As with all animals with eyes, humans have strengths, weaknesses and limitations in how they detect and translate light. Humans see better during day than at night, see a limited range of light and their eyes/mind do not translate light in an entirely efficient and accurate way. This all effects our visual perception.

Physical environment: This includes the brightness and angle of light, along with atmospheric conditions like fog, smog and air temperature. Many mirages are caused by unusual atmospheric conditions that distort light. In daily events the difference between light and dark, clear and cloudy can be the difference between identification and misidentification.

Biases: Humans have conscious and nonconscious, innate and learned biases that effect how they perceive. These biases are used to categorize, prioritize, label, translate and judge information. Biases often cause the viewer to perceive patterns, shapes, colors and identities that do not objectively exist. Biases cause us to place undue emphasis on trivial information, while ignoring what may be important. The proverbial missing the forest for the trees.

Personal knowledge and background: How a human perceives something is greatly affected by his knowledge, what he has been taught and past experiences. You identify a dog by having already seen animals and learned they are called dogs. Without that experience, you would be mystified by that strange creature sniffing around your neighbor's hedges. People new to geography often fall for

visual traps the natives do not.   In the crystal clear air of
mountains, long distance objects typically appear much
closer than reality. Newbies to high altitude are accustomed
to seeing through the hazier air of low altitude, with distance
of a far away object being judged in part by its relative
haziness—the further away a building or cliff, the hazier. In
an environment with clearer air, low altitude rules can
deceive. In highest altitudes a mountain can both be far
away and crystal clear, appearing closer than it is to the new
climber. As one might imagine, misjudging distance in the
mountains and cliffs can be dangerous.

Humans shape their perceptions and overcome many of
their visual illusions with experience. After walking into a
sliding glass door for the second time, you likely have
learned not to do it again. Humans aren't omniscient and
learn by trial and error, gaining knowledge as they go. In
many cases visual illusions are a natural part of the learning
process-- error in judgment (visual illusion), followed by
realizing the error, followed by having better knowledge.

(8)
Visual Illusions: Normal
Everyday Perceptions

reflection in store window   sun glare

There is perhaps no better place for a visual illusions discussion to start than with the visual perceptions in our normal, everyday lives. Our daily experience reveals how ordinary changes to our sight changes our perceptions of unchanging things. Many of these optical and mental tricks barely raise an eyebrow as they are commonplace and we know how they work. Not all of the following examples may qualify as illusions, but all reveal the variations in how we perceive.

* To read at night, you turn on the light. Though the printed words are identical in dark or light, humans can only read them in light. Without light the words disappear from human sight.

* If you enter your home and someone is sitting in a chair in your darkened living room, you may see the form of the

person but must turn on the light or ask to find out who it is.

* We often can't identify a bird or other small animal until we pick up the binoculars. It is with the distortion (magnification) of our eyesight that we learn the animal's identity. Without the distortion we can only guess.

* If there is enough sunlight glare on a downtown store window, you can't see what is past the window while standing outside. You can only guess what if anything is inside. Perhaps the store is filled with people and products. Perhaps the store went out of business last week and the inside is bare. When you walk up to the window and shade your eyes, the glare is removed and you see what is inside.

* Most shoppers have experienced where the color of the dress or paint or wallpaper or couch looked different when they got it home, caused by the difference in lighting between home and store. Some have thought they accidentally brought home the wrong item. Some have complained that the store intentionally used deceptive lighting.

* At night you are watching television and see something moving outside, only to discover it is a reflection of you in the window. You move your arms and head side to side to confirm it isn't a scary monster.

From these common situations you get a glimpse at how knowledge and perception are affected by even little things like light intensity and viewing angle. Even a slight change to viewing angle can be the difference between knowing and having to guess.

(9)
College

At a party the only thing I had with which to exchange phone numbers on was a pair of unused blue books from my backpack. I was about to ask around for something more normal to write on, but the woman found blue books amusing and insisted.

* * * *

My friends and I saw Seven Year Itch at Physics Hall where they showed old movies every Saturday night. We were surprised we'd never heard of Tommy Ewell before he as was the star of the show.

Authentic Colors?

1870s woodcut print of baseball team

1800s Harper's Woodcuts, or woodcuts prints from the magazine Harper's Weekly, are popularly collected today. The images show nineteenth century life, including celebrities, sports, US Presidents, war, high society, nature and street life. Though originally black and white, some of the prints have been hand colored over the years. As age is important to collectors, prints that were colored in the 1800s are more valuable than those colored recently. The problem is that modern ideas lead collectors to misdate the coloring.

Due to their ideas about the *old fashioned* Victorian era, most people assume that vintage 1800s coloring will be subtle, soft, pallid and conservative. However, 1800s coloring was typically bright, gaudy, bold and even tacky to modern taste. As Victorian people didn't have color televisions, motion pictures or video games, and were restricted in their travel, they liked their images of exotic places and faraway celebrities to be colored exciting.

A learned forger might knowingly use historically incorrect colors, as he knows the average person today would consider authentic colors to be fake.

(11)
When does 1 + 1 not equal 2?

Is a bag of potato chips one thing? Many? Both? Neither?
Other? Depends on how you look at it.

* * * *

A basic part of mathematics, physics, chemistry,
engineering, economics and daily life is counting. Counting
is popularly considered to be an objective activity. In the
field, however, it involves subjectivity. Not over whether 1
+ 1 = 2, but over what is 1. Both scientists and non-
scientists have personal and varying views of what is 1 and
what is 2, 3, 4 and 5.

Humans mentally, even nonconsciously, individualize
things, isolate, group and count things— whether or not the
things were designed to be individualized, isolated, grouped
and counted. To humans a dog is one thing. A cat is one. A
dog and a cat are two. This numbering is not just
intellectual, but often psychological, aesthetic, moral,
religious, political and philosophical. A human being is
popularly regarded as a single thing, a proverbial island unto
itself. Some will be morally offended if you count a human
differently. Two humans, even if physically connected by
holding hands, are not considered one human, but two.

A distant snow capped mountain of one billion stones is
commonly referred to as one thing, not one billion things.
Yet three of the stones removed and held in one's hand will

be labeled as three. This shift is a reflection of the counter's mind and eyesight more than the counted. Mountains and stones existed fine before humans were around to count and *individualize* them. How or whether or why we count them makes no difference to mountains and what they are. The counting is a human exercise.

When a long cloud briefly separates in the middle many call it two things, two clouds. What is the legitimacy of this representation? Could it just as well be called one? Is either number an arbitrary choice, a definition of terms?

A lake and connected creek that share the same water and fish are commonly considered two things. Is this the correct representation? Could they instead be considered one? Is 1, 2 or any number a true representation of the body of water, or merely a convenient representation for humans?

Perpendicularly intersecting roads are often considered two things, while a wooden cross is commonly considered one. What is interesting about this example is that the roads are more physically one than two boards nailed or glued together. If you stand at the middle of the intersection, the two roads at that point are physically the same. It is not one road or the other road, it is both roads simultaneously. A piece of asphalt belongs to both. The two cannot be separated or distinguished from each other. At the intersection of the cross, on the other hand, the two pieces of wood are easily distinguished and can be separated. Physically at least, the roads could be considered more one thing than the cross.

* * * *

My sandbox of stones

Say I have in my front yard a sand box filled deep with an unchanging amount of stones. Just as with a sandbox of sand, no matter how I fiddle or play or scoop or make stone castles there is never a gap with no stones.

In this ever unbroken sea of stones, I make two tall mounds of stones on the surface. If I pull someone off the sidewalk, point to the box of stones and say, "How many things do you see?," she likely will say two. She may even point out that the two things she sees are the mounds. If I had instead made three mounds, it's likely she will say there are three things. If there was one mound, it's fair to assume she would have said one. If the surface was flat (no mounds), she may say there is one thing. Even if her answers aren't as I just said, they likely would change depending on the number of mounds.

Duly note that my question was 'How many things do you see?' I didn't ask how many stones or how many shapes or how many mounds. I let the woman define what was a thing and count as she see fit.

There are two interesting aspects about her counting of things in the box. First, it is not clear that the number of things in the box ever changed. There was always a body containing an identical amount of stones. The body was constant, other than the changing surface shape. No one I know counts lakes by counting the number of surface waves. To most people, a strong wind doesn't create more lakes. People don't count triangles as objects differently than squares, or two humped camels differently than one hump camels ("Guess what, Mom. I saw two camels at the zoo today. One one-humped camel and one two-humped camel.") There was never any separation that created isolated islands of stones. It was the changing surface shape that caused different number answers. Her counting was personal. A different person looking at the same stones might come up with different numbers, as he defined *things* differently.

The second interesting thing was that, even if accepting her definition of surface mounds as the things, the woman's math was goofy. When there were one, two, three mounds, the woman counted things by the number of mounds. But when there were no mounds, she didn't say there was nothing. She likely would have said there was one thing (the body of stones) or been confused as to what she was supposed to count or perhaps said "There are a lot of stones. I can't count them all." Her definition of what is a thing and her method of counting was inconsistent. In her math, removing 1 thing (mound) from 1 thing did not equal 0, and in fact may have equaled more than 1.

* * * *

The act of counting the box of stones, or land or clouds or a herd of wildebeest, has at least as much to do with the counter, her biases and perceptions and idiosyncrasies and choices, as with the subject being counted. That the woman's definitions changed and different people off the sidewalk may count the box of stones differently demonstrates this. Many people believe that the individualizing and counting of things is intrinsic to the things being individualized and counted, but there is no evidence this is true. The human counting of a mountain may have nothing to do with what it is. Is a cross 1 or 2? Why does it have to be either?

Many will point out that counting is essential for humans, an important tool for functioning. This is correct, but again demonstrates that counting is about humans. Having a practical use doesn't make a conceit any less of a conceit.

Labeling via psychology. Even those who have never heard of the 1922 German silent movie *Nosferatu* form a strong psychological judgment of this still image. I don't have to tell anyone this isn't the tooth fairy. The movie's director was well aware what kind of feelings this image would evoke in movie goers.

(12)
Visual Illusions: What You See Is
Different Than What You Look At

Despite common belief, humans do not perceive a direct and exact representation of external reality, but a distorted translation formed by their eyes and mind. The image we see is different than what we are looking at. This is not some coffee house theory, but physiological fact. The human eyes and brain do a decent but imperfect job at detecting and translating light.

This chapter is a look at the physiology of seeing and offers examples of optical distortion caused by the eyes and mind. Sometimes this physiology is the primary cause for a visual illusion. Other times it is a lesser but contributing cause.

* * * *

A quick look at the physiology of seeing

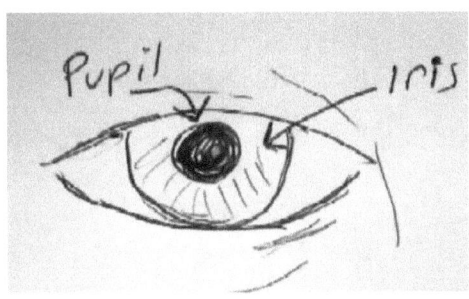

When a human looks at an object, light from the object enters the eyes. The light goes through the cornea, which is a clear covering, then through the pupil which is a clear circle in the center of the colored part of the eye called the iris. The pupil gets larger (dilates) when there is little light and smaller when there is more light. The lens focuses the light through the aqueous humor, a clear liquid, onto the retina. The retina, in the back of the eye, contains millions of tiny photo sensors that detect the light. There are two main kinds of photo sensors, called rods and cones. Shaped like rods, rods detect shades and forms and are needed for night and peripheral vision. Rods are not good at detecting color. Shaped like cones, cones are needed for seeing details, seeing in daylight and detecting colors. Cones do not work well in low light. Rods and cones cover the entire retina except for a spot where the optic nerve connects to the brain. The optic nerve carries the information received from the retina to the brain, where the brain translates it into the single image we perceive, or 'see.'

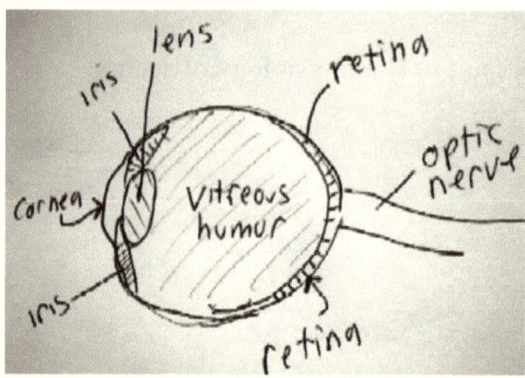

* * * *

The Blind Spot

All humans have blind spots, which are spots where the eye cannot see. The blind spot corresponds to the spot on the retina where the optical nerve connects the retina to the brain. At this spot there are no light detecting cells and, thus, it cannot detect light. A small object can disappear from view.

In everyday life the blind spot goes unnoticed. This is in part as the eye is constantly looking around, getting a wide and varied range of views. It is also in part as the brain uses the information from both eyes to create the single mental vision. What one eye misses, the other often picks up.

As its optical nerve connects differently, the octopus has no blind spot.

* * * *

Detecting your blind spot

L   **R**

To detect your blind spot using the above letters L and R, hold the book about two feet in front of your face, close your right eye and look at the letter R. Slowly move your head forward, towards the picture. At one point the L will disappear. The L will also disappear if you start up close and slowly move back. Notice that the missing spot is filled in white by your mind, so it appears as if nothing is missing from your view. This illustrates how your blind spot goes unnoticed during daily living. Many people live their entire life not knowing they have a blind spot.

* * * *

Humans have more glaring blind spots. Due to the placement of our eyes, we can't naturally see behind us, under our feet, from the top of our head, behind our elbows. A common saying to explain why we didn't notice something is, "I don't have eyes in the back of my head." If you want to sneak up on a person you approach from behind. We compensate for these blinds spots by turning around, moving our heads, listening, noticing shadows, saying "Who's that behind me?"

Other animals have different eye placement and fields of view. As a robin has its eyes on the side of its head, it has better side view than humans, but worse direct ahead view. The robin's life depends on its being able to detect predators from the side and back. When hunting for worms in the grass, robins turn their heads. Some people think they are listening for worms, when they are turning their heads to see in front of them. A wolf, who is a hunter stalking prey, has eye placement best suited for seeing ahead. With eyes in the front of its head the wolf sees better straight ahead, but its side to side vision is worse than a robin's. A crocodile has eyes that rise above the rest of its head. Not only does this create a different field of view, but gives the crocodile periscopes to see above water while the most of its head and body are hidden below.

* * * *

After Images

Afterimages are when, after staring at an object, you look away and still see an image of the object. An example is when you still see the nighttime headlights of a car, even though your eyes have closed and the car has turned away. Another is when after looking away from a candle flame in the dark you still see light in the shape of the candle flame.

Afterimages happen after the retina's photosensors (the rods and cones in your eyes) become oversaturated, or burned out, from staring at a particular color. This burning out is comparable to lifting weights in the weight room. After doing enough arm curls you lose your arm curl strength for a short while and will be able to lift only lighter weights. Your muscles are fatigued, if only temporarily, from all that weight lifting.

Similarly, after staring at a large area of a single color, the eye's photosensors lose their strength for that color. If right afterwards the eyes look at a blank piece of paper, the photosensors will be weak towards the previously stared at color but fresh and strong for detecting the other colors. This imbalance causes the mind to perceive the image (the afterimage), but in the color opposite to the original color. To the mind, the weakness towards one color means the presence of the opposite primary color is stronger. Quirky perhaps, but this is the way the brain works.

If you are staring at a green image, the afterimage should be red (the opposite primary color). After staring at a yellow image, the afterimage should be blue. The mind sees afterimages in primary colors, so any non-primary color (orange, pink, etc) will be seen as the primary opposite. The below chart shows you what the opposite primary color would be.

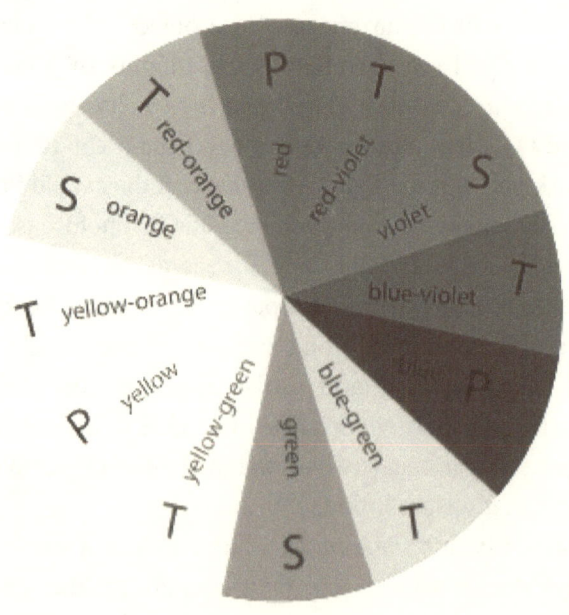

Though they occur almost constantly, afterimages usually go unnoticed. Afterimages are best observed when focusing on a single color or object for a lengthy period of time. In normal about the house viewing we view a wide range of objects and colors at once and our eyes are always moving around, the view constantly shifting. In these cases, the afterimages are minor and get lost in the visual shuffle. We barely if at all notice them.

* * * *

Natural delay in processing light

If in the dark you quickly pass a lit match in front of your face you will see a trail of light following the match. If you pass you hand quickly in front of your eyes in daylight your hand will blur. This effect happens in part because your eyes and brain don't process light instantly. It takes a small fraction of a second for the eyes and mind to translate the light that enters the eyes into the mental image you see. Your hand passed quickly by your eyes looks blurry as it is moving faster than your eyes and mind can process. This explains why humans can't see speeding bullets.

* * * *

Binocular Vision

Humans have binocular vision, meaning the single image we see in our mind is made from two different views-- one from each eye.

Binocular vision gives humans a number of advantages. One is we have a wider field of view than if we had only one eye. The right eye can see further to the right and the left eye further to the left. The single vision in our mind shows more than either single eye can see.

Another advantage is the two views give us imperfect but good depth perception. People who are blind in one eye have worse depth perception than the average human.

The mythical Cyclops might at first appear an unbeatable foe, but a wily human opponent could take advantage of the monster's poor depth perception and narrow field of vision.

* * * *

Triangularism and Calculating Depth

Binocular vision produces the perception of depth in a way similar to how triangularism measures length in applied mathematics. When looking at a distant point using only one point of view it is hard to impossible to determine the distance accurately. In applied mathematics, triangularism can accurately calculate this distance from point **a** to point **b** by creating an imaginary triangle. Trianglularism has long been used in the real world to measure distant objects, like islands and boats from land and when surveying land.

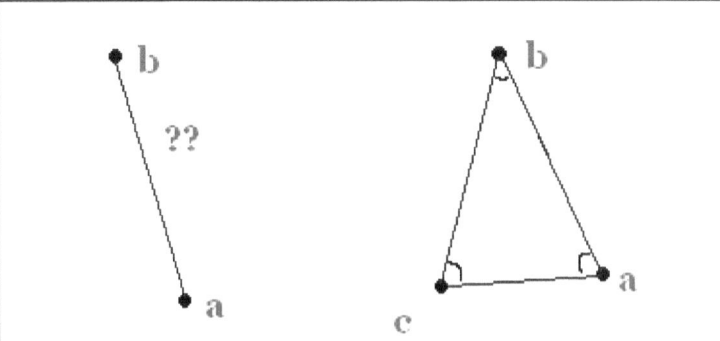

Triangularism: From point a alone, it can be impossible to accurately calculate the distance to point b. In the real world, point a could be you standing on land and point b an anchored boat out at sea. However, by taking angle measurements from point a, then taking an angle measurement from nearby point c (perhaps a walking distance away), and measuring the distance from point a to c, one can create an imaginary triangle that calculates the distance from point a to point b. It's just a matter calculating angles and doing the math.

Two eyes give the mind a similar two point view, and the mind uses these two views to judge distance. This is mostly done nonconsciously. You simply reach out and grab that pencil or door knob, no problem. If you wear an eye patch, you may discover it's more difficult to grab things on the first try.

The Hole In The Hand Illusion

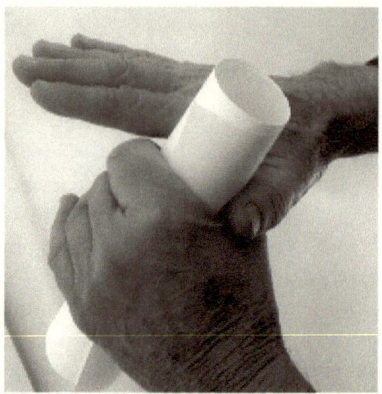

This simple trick plays with your binocular vision to make it appear as if you have a hole in your hand.

Roll a normal piece of 8x11" paper into a tube and place it next to your hand as shown in the above picture. With one eye look through the tube and with the other eye look ahead at the back of your hand. With a little bit of shifting you should see what appears to be a large hole through your hand. Your mind takes the two distinct views to create one bizarre view.

(13)
Visual Illusions: Night Versus Day Perception

Even when the objects remain constant and stationary, humans perceive them differently in light versus dark. To humans, objects "appear in the light" and "disappear into the dark."

Due to the physiology of our eyes and how the rods and cones detect light, humans physically see better in daylight than in dark. Humans see more detail and color during daylight. At night things become murkier to hidden, colors fade.

There almost always is light when it is pitch black to humans, but it is in wavelengths human eyes can't detect. Ultraviolet and infrared light are commonly present, but invisible to humans. A human can get a suntan from ultraviolet light and feel the warmth associated with infrared light, yet is unable to see either.

Other animals have night and day vision different from humans. Owls see better in night than in day. It's not that objects like picnic tables and fence posts physically vanish in the dark of night. It's that humans are unable to see them. Owls see them fine.

Geese see ultraviolet light invisible to humans. Geese eyes see all the color we see, plus the color of ultraviolet.

* * * *

Human thought and culture are shaped by their optical abilities. The night is mysterious with information hidden from view. It is difficult for humans to function at night, more dangerous and filled with out-of-the-ordinary nocturnal creatures. To humans, the night is another world, a place to be avoided or entered with caution.

From a practical standpoint, wariness of night makes sense. For example, it's safer for you or me to sprint through the woods during day than night. That's not superstition, that's good sense.

Darkness is popularly associated with sinister, and light with goodness. Look at the common dark words and phrases:

Dark angel
Dark and mysterious
Shadowy figure
Dark thoughts
Dark and sinister
He has a black heart
The darkness of his soul
Dark motives
He has a dark mind
Heart of Darkness

Human society mostly functions during daylight hours. Elementary schools don't run at night.

The color black is worn as a statement by brooding teenagers.

In Western culture, white, yellow and other bright colors are associated with happiness and goodness. Someone who is upbeat and smiley is said to be in a bright or sunny mood.

Hell is commonly pictured as shadowy and Heaven as sunny. Good angels are typically described as wearing white. Virginal brides wear white. The Wicked Witch of the West wore black. The Good Witch of the East wore white.

Monsters are commonly called creatures of the night, and genuine creatures of the night, like bats and owls, have been called monsters and demons.

Vampires, as the stories go, rise at night from their coffins and die when exposed to daylight. The cursed man becomes a werewolf at the full moon of the night.

* * * *

Off Center Night Vision

Have you ever noticed that when you're outside at night, you can see a star better when you're not looking straight at it? The center of your retina does not have rods which are used to see at night. The rods are off center, so you see better at night off center. When looking at a faint star, try turning your head a bit as it may appear brighter out of your periphery.

* * * *

Ghosts

Given humans' night vision it is not coincidence that humans perceive ghosts as things that come out at night, are pale and colorless, ephemeral, fleeting, peripheral, dreamlike, shimmering, mysterious, otherworldly. Under the shroud of night a lawn chair can look like a shadowy figure. A backpack left on a picnic table can resemble a strange animal. A rustling bush can startle the sheckles out of someone walking home alone. As there is a lack of visual

information at night, humans use their imaginations to fill in
the story.

Infrared viewers, such as night vision goggles, do not allow humans to see infrared light, but translate infrared light into visible light. We cannot see infrared light and can only guess how an infrared viewing animal perceives it.

(14)
Visual Illusions: Pattern, Shape and Form Biases

Human visual perception is profoundly influenced by biases about forms, shapes and patterns. Humans have ingrained and nonconscious attractions for specific forms, shapes and patterns. Some of these biases are genetic, while others are learned. These biases greatly influence how we perceive, organize and label, and are essential to the quick identifications needed to go through life.

Humans can naturally tell the difference between a perfect and lopsided circle and between a straight and slightly bent metal bar. This ability is cross cultural. Someone in Berlin and someone in Cairo have the same ability.

From having looked at pictures in books, magazines and on television, a kid can immediately identify the distant form of a Grizzly in the wild, even when the bear is silhouetted by shade. The same kid at grandma's can instantly identify a cookie by its gingerbread man shape.

You instantly perceived a dog in the black shape that started this chapter, even though the shape lacked fur, eyes, whiskers, correct size and other essential doggy details. You didn't have to contemplate the shape. You perceived it instantly.

The problem for humans is that their biases for certain shapes, forms and patterns are so strong and ingrained that they will perceive these things when don't objectively exist. These biases lead to many visual illusions.

* * * *

Creatures in clouds and other personalized visions

Our form and pattern biases are shown when we perceive horses or castles or hot rods or other familiar shapes in clouds. These 'identifications' are subjective to the viewer, and do not objectively exist in cloud. There are thousands of possible connect-the-dot shapes in a cloud, but you perceive, or mentally pick out, that which matches your knowledge. The horse or castle is a projection of what exists in your mind. If there were no horses on earth or in fantasy books, you would not perceive a horse in the cloud, as you wouldn't 'recognize' it.

The connect-the-dot figures in stars don't exist except as we draw them. The familiar faces or figures we perceive in burnt toast, wood grain and stones are projections of our minds. What you perceive is as much a reflection of you as what you are looking at.

I hope it dawns on you when you pick up a stone that 'looks just like Elvis,' the stone existed long before Elvis was born. It would be silly to believe the stone was formed by glaciers 10,000 years ago to commemorate Elvis' future rise to popularity on the pop charts.

* * * *

The Face on Mars

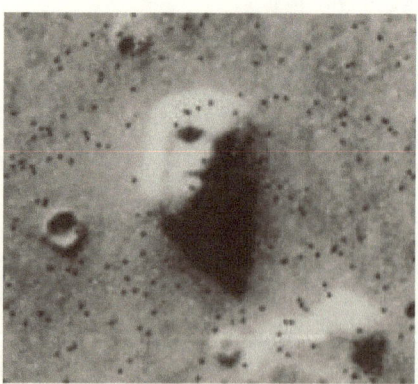

In 1976 the NASA spacecraft Viking 1 took photographs of an area on the planet Mars that contained many giant mesas, craters and other geological formations. One of the mesas in the photographs somewhat resembled a human face (see above). As should not be unexpected, many humans on earth became interested in this 'human face' (and, not surprising, were less interested in the formations that didn't resemble human body parts). Some were and still are convinced the mesa was constructed by intelligent life form.

This perception of a face is a pattern bias, a projection of the viewer's mind whose own face has a similar form. If someone has patterns in his mind (human face, kitty cat, square, letter 'B,' house key, baseball cap, house) and looks at enough information (such as all the geological formations on a planet's surface), he will be able to pick out some of these patterns in the information.   Seeing the 'face of Nixon' isn't proof a potato was built by intelligent life form. It means that out of millions and millions and millions of potatoes, a few are bound to somewhat resemble a former US President who had a sticky outy nose.

As the following images show, the face on Mars is just one of many mesas, hills and craters that come in a wide variety of shapes.

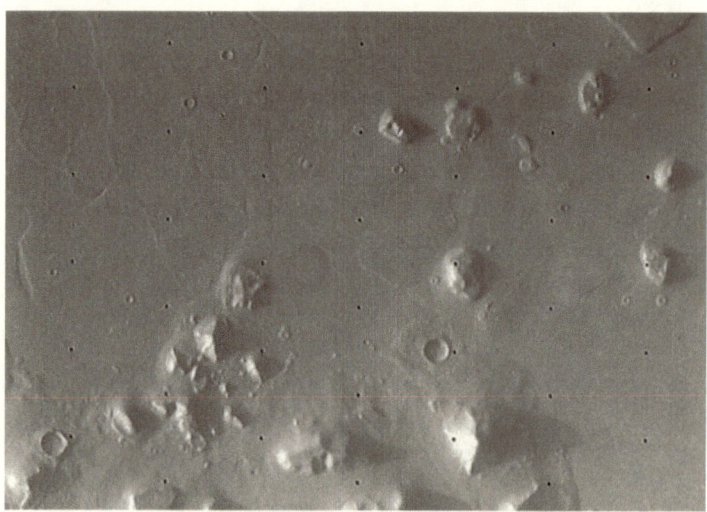

Just another mesa in the crowd

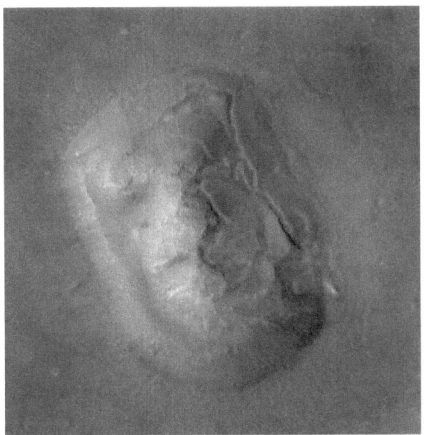

Years later, the above photograph of the same mesa was shot at a different angle and time of day. This shows that angle and shadow contributed to the perception of a face. If originally shot at this angle and time of day, the mesa may not have been perceived as a face and humans on earth would have considered it no more significant than any of the other blobs in the photographs.

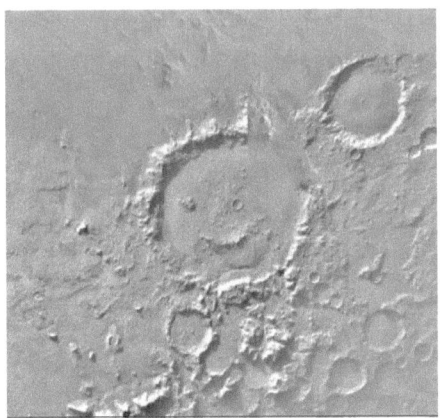

Smiley face of Mars?

* * * *

```
/////////////*/////////*        ***//***//////////
//////////***/////////          /////////////////////
/////////****///////            /////////*******///
/**////******//////*            ////////****/////////
//////********///                ******/*/****///
 //////*********///             /////////*/*/*/*///
 ////**********//               /////////////////////
////**********///               //////*****//////
///////////**/////////          /////*//*//////////*//
////////*/***//////////          */*////////////////**
```

When most people look at the above two randomly ordered designs, they single out the left image as more significant as it contains a pattern they recognize— 'a tree.' Most will label the left design as being more ordered and the right design as being disordered or unfinished.

* * * *

Is it a Vase or Two Faces?

The standard *Is it a vase or is it two faces looking at each other?* visual illusion shows that humans project a subjective, or personal, identity onto an object. You initially see a black vase or a pair of white faces looking at each other. As you stare longer your perception will be replaced by the other view, then your perception will flip back and forth between the two views. The image is unchanging, while your perception of it changes.

Of course it is neither a vase or faces, but a black and white abstract pattern. The pattern could be perceived as many things. However, in part by your biases and by the leading question ('Is it a vase or faces?'), you perceived a vase and faces. As I look at the image, I could see how the top or bottom portion could be perceived as two boots placed back to back. The chin to nose areas could be perceived as little black faces. The black shape could be seen as a table. In fact there's no reason, beyond viewer's predilection for order, that the pattern has to depict anything specific.

15

16

17

(18)
Monster

We cannot enter into the human world-- the daylight world--
though we can see them, meet them at night-- I have talked
to a few-- observed them

At times I've walked in the day-- just a few times-- Met
some of the daylight ones-- An old man on a bench who
talked about baseball

He walks along the sidewalk and doesn't know what to do.
He begs the toad, he begs the grass but they don't talk back

I can't really describe myself (me = do, me = iguana)-- I'm
mostly light, ghost sort of
I have flesh but it's like flesh in a dream-- you can touch
it/can't touch it

The beloved Birgit grows old, dies- SH continually dates
young women-- who shrivel up and die

He says he is the only one of his kind on earth-- with nothing
to do--
As he grew out of being a human being-- he lost their desires
for day to day, light, etc-- And his temper, desires grew--
Can only think big

Finds spy-- confronts spy--- Spy says he doesn't know what
SH is talking about

I must paint scenes-- I must promote scenes-- I must provide
the music and food and drink for the scenes. And I must
enter into the scenes. I must beg of them, I must plead of
them

I am forced to be a human, etc.-- have a wife and daughter,
etc--
We travel in a car
I can go along with being a human-- but there is an
uneasiness-- the busting--
it grows larger-- I can no longer drink their water

I watch a beautiful, painfully boring couple sitting on a
bench.

I made a machine to put the image onto paper-- To have
something tangible-- But this didn't help-- I held it, caressed
it-- I made it into a paste and injected it into my vein-- But it
just made me sick

Then in desperation he goes up to a woman—She doesn't
know what he's talking about

(19)
Visual Illusions: Comparisons

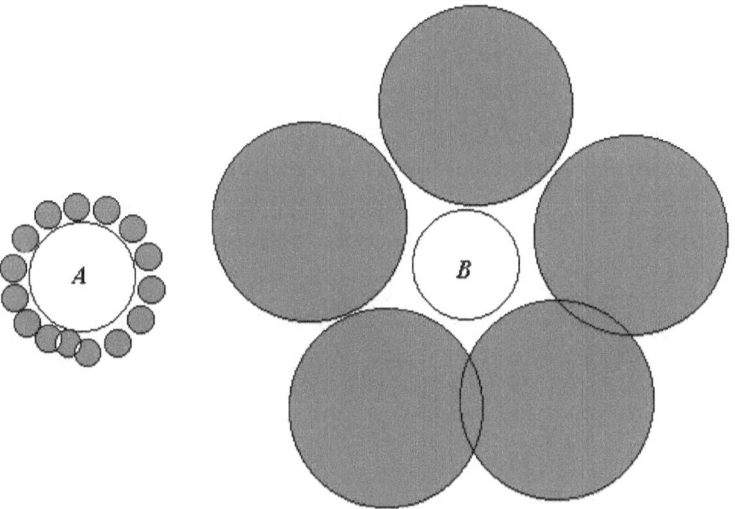

Circles A and B are the same size. It the surrounding grey circles that make circle B appear smaller.

* * * *

Human perception of objects is influenced by nearby objects, qualities and other information. Both consciously and nonconsciously we judge things through comparison. To measure fabric one compares the cloth to a yard stick. To judge the size of someone's hand, you might press your palm against hers. To judge someone's speed, you might race him or watch him race someone else.

In often less exacting comparisons, humans judge the length, height, angle, shape, color and distance by comparing one object to others in the scene. Looking at a family snapshot photo you might guess the height of a stranger by comparing him to someone you know. You will guesstimate the distance to a house by comparing its size to the sizes of closer houses and trees. You will guesstimate an angle by comparing it to a level line ("Appears to be about 10-15% off from level").

Often these guesstimates are accurate within a reasonable degree. You might guess that stranger in the photo is 6 feet tall, as you know your aunt is 5' 5." When you meet him, you may discover he's 5'10-1/2." Not perfect, but a darn good guess—especially as you were unable to clearly see what shoes they had been wearing.

A problem is that, while comparing to other objects is essential to making judgments, comparisons can lead to errors. Seemingly logical comparisons can produce answers that are bizarrely wrong. These errors happen when assumptions about an object or about the overall scene is wrong.

What happens if you incorrectly remembered your aunt as 6 feet tall, instead of 5'5," as the last time you saw her you were a five year old munchkin? Your calculations of the man's height will be similarly off. You might wrongly guess he was 6'7." What happens if she was wearing flats in the photo, while he, shy about his height, was wearing lifts? What happens if the man couldn't make the family reunion and a cousin photo-shopped in an image of him?

The following pictures show how your perception is distorted by surrounding information.

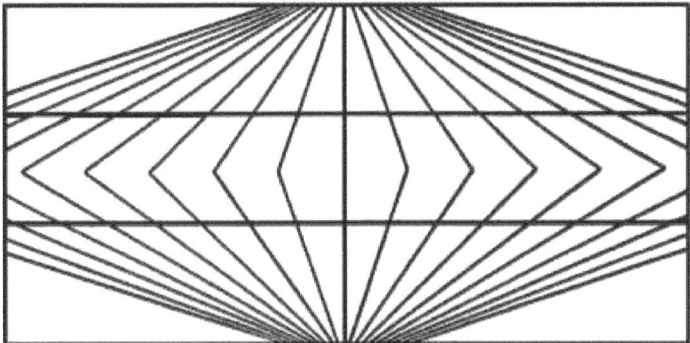

The above two horizontal lines are straight and parallel. The angled background makes them appear to bend. Without the angled background, the lines would appear parallel.

All the horizontal and vertical lines are straight and equally spaced. In other words, all the large checkerboard squares are perfect squares of the same size. It is the placement of the tiny squares that creates the appearance of the 'bulge.'

The men are the same size. Measure them yourself. It is the skewed *diminishing scale* lines that make them appear to be of different sizes.

'Gravity Hill'

A so-called gravity hill is a visual illusion where a road appears to go uphill when it actually goes downhill. A car left in neutral appears to do the impossible of rolling uphill. There are gravity hills in many places around the world.

These illusions happen in areas where the road slopes only slightly and often where there is an unusually high horizon (hill, quarry backdrop, other). Humans use the horizon line to judge the slope. If you look at a road moving away from you and steeply upwards, it visually moves towards the horizon. If instead it is moving steeply downhill, the hill moves away from the horizon. With a slight slope and unusually high horizon, judgment becomes more difficult. Additionally, gravity hill areas often have unorthodox angled scenery, including few if any perpendicular objects like trees or telephone poles. These perpendicular objects also serve as references for judging slope.

While these roads are visual illusions, locals often refer to them as them as being supernatural, haunted, magnetic, gravity or anti-gravity hills. These local legend terms are tongue in cheek to have some fun with tourists.

The road on the left clearly moves up, moving towards the sky. The vertical phone poles going up the hill aid our judgment. The slope of the road on the right is harder to judge. This is due to the lesser grade and the unorthodox, almost bizarre angles of the scenery.

Which cyclist is going fastest? Most will say the cyclist on our left is going the fastest and the one on the right the slowest. There are, however, unanswered questions that make it impossible to know. Did they start at the same place? Did they start at the same time? Are they moving forward or backward? Are they moving? (I've seen sprint cyclists stand still during a race.) Even if it's a normal *1-2-3-Go!* race, it's possible the guy on the right is going the fastest and the guy on the left the slowest at the moment the image was shot.   Catching up, slowing down and switching positions are normal parts of all races.

Camouflage : Seeing But Not Perceiving

With some forms of camouflage, like a brown chameleon standing in front of a matching brown rock, you see the chameleon but don't perceive it. Your eyes and mind receive the same visual chameleon information as when the chameleon is standing in front of a white sheet.   Humans can't visually identify anything without contrast. This is how a chameleon or arctic fox can hide in open view.

(20)
Fiction in Science

Scientific representations are different than the things they represent. A representation, model or description is a limited view of the subject, made for a specific purpose, edited by the scientist and translated into a form the scientific audience can understand and use. As scientific representations are made by and for humans, they are part about the scientific subject and part about the humans using them.

* * * *

A world map is a useful device, but one with a plethora of differences than what it represents. To start with the obvious, the world isn't flat and it isn't paper thin. These unreal qualities are for the convenience of the user.

For easy understanding, maps are artificially colored and marked (latitude and longitudes lines, for example). Road maps usually make roads appear proportionally wider than in reality, and remove unwanted details.

All world maps have proportional distortions. For an example see the map on the following page. Translating anything three dimensional into two dimensions requires distortions, as three dimensions and two dimensions are mutually exclusive. Compare your world map at home to a globe and see the differences for yourself. There are different methods of mapping the earth, each method creating its own distortions.

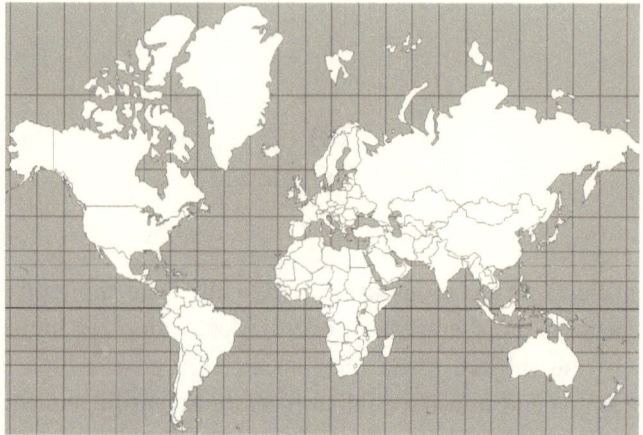

Distortions on maps. As with all types of world maps, this common Mercator projection map has significant distortions. Greenland is incorrectly shown as being bigger than Africa. Alaska is shown as being as large as Brazil, when Brazil is really multiple times larger.

* * * *

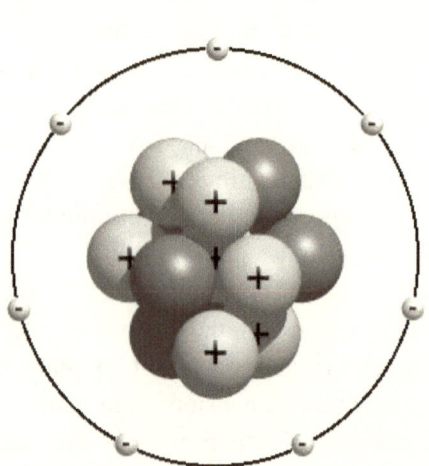

The above representation of an atom is different than a real atom in an abundance of major ways. To start, it's thousands and thousands of times larger than a real atom. If it wasn't you couldn't see it.

The representation hardy resembles an atom, and the artist would agree. The intent was to make a dummy model for students to learn about the different atomic 'parts.' The unreal balls, outer ring and cartoonish appearance are designed to engage the audience, simplify things.

As with the map, this representation is part about the subject and part about humans. It is in a form students can understand. In this case the form students understand looks more like a Saturday morning cartoon character than an atom.

* * * *

Any human representation of something complex (and all things are complex) is simplified and distorted, focusing on a specific area, quality, layer or angle, made from a limited amount of information, interpreted by the maker's sensibilities, presented in a way the maker and audience can understand.

As a means of communication, a representation will include conceits of the scientist, audience and even general culture. These conceits include expected form (pie charts, graphs, book, magazine article), style, shape, measurement method (volume, height, meters, liters), color associations (hot = red, cold = blue, forest = green).

It is similar to art, where following the genre's conceits, even shallow ones, are constraining but necessary for communication. The conceits create an artificial representation, but without them you might as well be communicating in a foreign language.

* * * *

Just as the creation and perception of art involves human psychology, so does the creation and perception of scientific representations.

Whether they admit it or not, scientists and philosophers view the universe and the things in it psychologically. A scientist and his work can no more escape human psychology than the scientist can escape being human.

* * * *

All one has to do is to look at a scientific representation, any representation, and find the human imprint— the human sensibility in form, style, color, language, balance, aesthetic choice. A representation of water may be a magazine article in English. English language and magazine articles, of course, have to do with humans and communication between humans. The article's subject may be about water, but its form is human. The article will be read as a work of human literature, as it is a work of human literature. As an artifact, the article shows about as much about humans as it does about water.

* * * *

Organize the following into two groups of related objects

Scientists, and non-scientists, often find it convenient and practical to group information. I asked different people, including a science professor, to group the above objects into two groups of like objects. One person grouped by color (black objects and white objects), another by size, another by letters (he saw the objects as E's and C's. Interesting, as I drew the Cs as moons!), another by direction left or right (problematic as one doesn't know if a moon is faced left or right). Their reasons for pairing were equally legitimate, but produced different pairings. This should show you how one scientist's model can look different than another's, not due to scientific theory or knowledge but different views of aesthetics, simplicity and association.

* * * *

This illustrates an essential human problem that goes beyond science. Humans must translate a subject to understand it, but what they understand is the translation.

* * * *

A scientific representation is a product of the scientist's purpose. A different purpose will produce a different representation of the same subject.

I own three maps of North America. One represents the altitude (mountains, valleys, etc), one shows the traditional aboriginal tribal regions and one is a road map. Even though they are of the identical place, each map is different. It's not so much whether the maps are right or wrong, but that they were created from different purposes.

* * * *

Many to most scientific representations aren't intended to be the *be all and end all*. Scientists usually consider scientific models to be works in progress, to be studied, tested, reworked, changed and even tossed aside as necessary. Science is a continual work in progress.

For testing purposes, models are often intentionally made to be overly simple. One purpose of such simplification is that errors are more easily identified and corrected. With a more complicated, muddled model, it's harder to identify what is working and what is not. Another reason for simplification is the scientist may be studying only one aspect of the subject. The other aspects are excluded. If a dentist is studying the teeth and gums, there may be no need for her computer model to be full-bodied, including detailed feet, fingernails, hair color and bellybutton. It may not even include eyes and nose, even though people with teeth and gums also have eyes and noses nearby. She may consider these details distracting and "beside the point." A scientist will often be the first to say his representation isn't a duplication of the subject, and was never intended to be an exact duplication of the subject.

As with communicating of scientific ideas to others, reducing a subject into a simplified if unrealistic model has practical purposes. Scientific progress would be stunted without simple, artificial models.

* * * *

Knowing that all representations contain fiction, a question to ask about a particular representation is whether the fiction is a device required for communication of ideas, testing or other practical use, or is it wrongly portrayed as part of the subject's innate meaning. If you are well aware a fiction is fiction, there is no big issue. If you confuse fiction for fact, that is a problem.

While fiction, the size of the earlier representation of an

atom is needed for humans to see the representation. If the representation was life size, it would useless to instructors and students. Similarly, artificial color coding for a diagram or map can make for easier and quicker understanding. It's easier to find countries on a map if each is distinctly colored. These are examples of where the inclusion of artifice is fair and understandable.

A related question is how seriously is the fiction taken, both by the creator and the audience. Students and even seasoned scientists can become too comfortable, too enamored with clichés of color, shape and words. Through repetition, superficial conceits can become false idols.

hidden information and identification

Many physical qualities of and between objects are identified by obscured or otherwise hidden visual information. Distance is in part judged by objects overlapping each other and things becoming harder to see over distance. Material is in part identified by its opacity.

Thus, obscured visual information both helps and hinders our identification. When something is very far away, the lack of detail (very small and blurry) serves to both help us judge its distance and prevents us from identifying the object. A closed closet door shows us that the door is in front of the things in the closet, but prevents us from knowing what are those things. Lack of information is both lack of information and information.

That some parts of this table visually obscure other parts helps the viewer identify it as at table.

Overlapping helps show us that the scrambled eggs are on the plate and the trees are closer than the buildings.

(21)
The Unique Subjective Experience

Subjectivity is a constant and integral part of the human experience. Love, lust, like, dislike, taste, smell, views about beauty and ugliness and art. How you view this paragraph and this book involves subjectivity— your taste about the writing style, word choice, chapter subjects and length, book cover.

By definition, a subjective experience is a product of the individual's mind. While real and often profound, the subjective experience cannot be objectively measured by others. When someone is listening to music, the music's note, pitch, speed, volume and the listener's ear vibration and heartbeat can be measured by scientific instruments, but the listener's aesthetic experience cannot. This experience is experienced by the listener alone. Even if asked to, the listener could not fully translate the experience to others, in part because it is beyond words.

It's doubtful that two people have the same subjective perceptions. People may have similar, but not identical perceptions. People regularly like the same song but perceive it differently. It's common for best friends to like a movie, but one likes it more than the other or for different reasons.

* * * *

A large range of things determines a person's subjective perception and experience. This includes genes, education, culture, where and when born, personal experiences, upbringing, travel, family make up and personalities, friends, acquaintances, natural temperament, mental abilities, physiological abilities (quality of eyesight, hearing, smell), talents, language, health, hobbies and work.

Little things influence, like what toy one had as a six year old and what tea grandmother drank. While walking in a foreign land, the scent of jasmine tea can bring back a rush of memories. The appearance of the toy in a movie will alter one's emotional reaction and interpretation of the move. It may have been chance that the movie viewer's parents bought that toy, making his movie interpretation a result of chance. It's not just the tea and a toy, but millions of little things that influence, including from forgotten events.

If a bird watcher and a rock collector go for a walk together in the park they may have equally grand times, one due to the birds in the trees and the other due to the rocks on the ground. Though they were side by side, they will give decidedly different descriptions of the walk.

Do you dislike a name simply because it was the name of someone you couldn't stand?

* * * *

Even when they experience similar feelings people will usually have these feelings under different circumstances, if only slightly different. People will be artistically excited, but for different works of art or when interpreting differently the same work of art. People have similar feelings of romantic love, but for distinctly different people— different looks, personality, culture, interests, sex, race. The emotional states may be alike, but the objects of desire are not.

* * * *

You cannot separate your biases from your perception, because it is those biases that help create the perception. Without those biases, you would have a different perception. Even that childhood toy affected the movie goer's perception thirty years later.

* * * *

Humans believe they receive important objective insights, including cosmic truths, through strong subjective experiences— such as through the sublime experience of art, epiphany of music, nature, love, lust, religious experience. The psychological power of these experiences is considered verification of the 'truths.'

A question is whether these experiences involve genuine insight into external reality or are merely strong biological reactions. Love and lust themselves, after all, are standard genetic reactions. Psychological reactions to certain sounds, such as in powerful music, involve genetics.

The reactions to high delicate notes (like from song birds or a pop song) and low booming notes (distant thunder, the start of Beethoven's Fifth Symphony) have been shared by humans for thousands and thousands of years. You and your ancient ancestor have remarkably similar psychological reactions to the sound of a songbird and the sudden deep roar of a bear. It's not coincidence that church music uses delicate high notes to invoke heaven in the audience, and the loud, deep bass of the organ to invoke power and awe.

It's not coincidence that horror movies use discordant notes. The director knows audiences find the sounds scary and creepy. In the famous 1960 Psycho shower scene, the sharp, grating, discordant musical notes invoke violence, evil, something gone horribly wrong. They sound similar to someone scratching a chalkboard, one of the most despised sounds to humans.

It can never be known to the experiencer that an epiphany made through a strong psychological experience is anything more than a genetic reaction. If there is insight into the external, the insight is shaped by the expieriencer's subjectivity, and what parts of the insight are objective and what parts subjective is unknowable.

Even if important insights into the universe are gained they still are in subjective format. For example, if your epiphany comes through your experience of art, your experience of art is personal and different than that of others. Not only is your 'insight' intrinsically tied to your subjective views, you likely would not have had the insight at that same time, place or format, or at all, if you had different aesthetic views.

* * * *

Humans use aesthetic rules for defining truths, including what is good and evil, what is moral and immoral. Common rules include conditions of beauty, symmetry, color, tone (light versus dark), fashion and order.

Even if the rules were valid, it would mean truth is subjective. If truth is beautiful, your definition of what is beautiful differs from others' definitions. Further, an individual's perception of beauty changes with time and experience. A culture's perception of beauty changes with time. Compare the depictions of the desirable feminine body from 1450, 1850, 1950 and this year.

Cultural definitions of 'objective truth' are formed by cultural sensibilities, including fashion, politics, gender, race, beauty, geography, self interest, desire for social order, etc. There is no indication these are identifiers of objective truth, or are even related, but they are still used as criterion.

Mirror image

A mirror mirrors what is in front of it. If you place an apple two feet in front of a mirror, an identical looking apple will look as if it's the same distance behind, or into, the mirror's surface. Curiously, if you use mathematical triangulation to measure the distance to the apple in the mirror, the apple will measure as being two feet behind the mirror. Both our eyes and scientific measurement say there is an apple two feet behind the mirror's surface.

Instant perception

Humans make many visual perceptions in an instant. In an instant can mean the instant eyes are laid upon a scene. It also sometimes means the perception suddenly flashes in the mind after looking at the scene for a while. An example of the latter is when you stare at a Magic Eye picture before the hidden image is suddenly revealed.   Another is when you are looking at a crowd of faces and all of a sudden recognize a friend.

These instant perceptions often come from the nonconscious. That they arise instantly and from within, like epiphanies, make them powerful, even when wrong. To true believers, they didn't arbitrarily pick out the Face on Mars, they recognized it as one recognizes a relative in a crowd.

(22)
Visual Illusions: Imagination

When looking at a scene, all humans have the natural and nonconscious ability to extrapolate beyond what is visible. When information is missing, or assumed to be missing, humans make it up in their minds.

This ability is essential to normal living, as we must regularly make quick guesses with limited information. When you step on a sturdy looking building step, you assume it will hold your weight. When you pull a book from the library shelf, you assume the pages are filled with words. When your waitress brings you a steaming mug, you assume it is filled with a hot liquid.

In many cases the extrapolation is accurate, or at least a fair estimate of reality. If your dog is standing on the other side of the open doorway, half hidden by the wall, you correctly assume an entire dog exists. As the dog steps forward into the room, your assumption is proven correct. When the waitress puts down your steaming coffee mug, you are far from surprised to see it's filled with the hot coffee you ordered.   Humans would be a dim, slow species if we couldn't make these kinds of elemental deductions.

In many cases, however, the extrapolations are wrong. These bogus extrapolations involving the viewer nonconsciously perceiving what he wants to see or expects to see.

The following pages show examples of correct and incorrect perceptions based on imagining what is not seen.

Though the dogs block our view we assume there is snow behind them like the snow we see surrounding them. This assumption is likely correct.

Though the overlapping prevents us from ever knowing, most will assume the above shows whole playing cards. I assume the cards are rectangular and whole.

The below says *I Love You* several times:

Now read the same text below with the ruler removed:

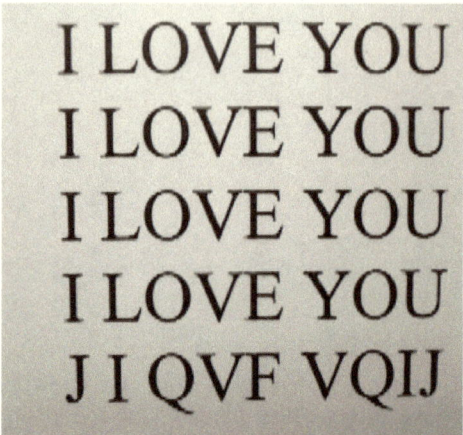

I LOVE YOU
I LOVE YOU
I LOVE YOU
I LOVE YOU
J I QVF VQIJ

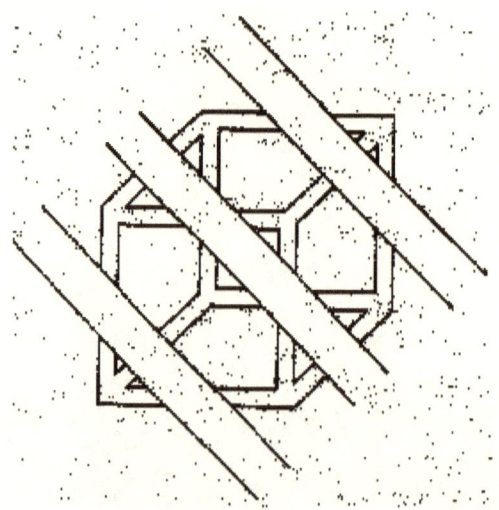

In the above, most perceive a cube behind the three diagonal bands

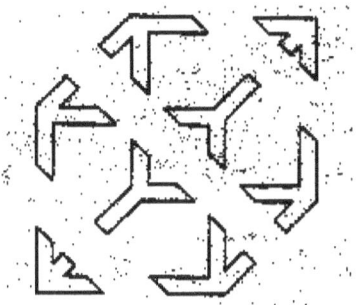

With the bands removed, we perceive something different.

* * * *

Ames Card Trick

Adelbert Ames Jr (1880-1955) was a Dartmouth scientist famed for his studies of visual perception. From his experiments he learned that human perception is influenced by expectations. In one experiment he used a deck of playing cards standard except in size. Some cards were much larger than the others. In a specially designed viewing station he had people watch the cards being shuffled. As the viewers assumed the cards were the same size, they perceived the larger cards as being closer and the smaller cards as being further away. Their misperception about size created a second misperception about distance.

(23)
Subjective Categorization, Grouping
And Prioritizing of Information

When a human being visually perceives, she mentally organizes, sorts, groups, prioritizes and labels the things in the scene. When you look at an ink sketch, you mentally assemble the ink lines, squiggles and dots into a form. "It's a kitty cat." "It's a cottage in the woods." You decide which ink marks belong together and how, and which do not. Two people can and do group the ink markings differently.

A Rorschach ink blot is perceived differently by different people. The ink blot remains the same. The viewer changes. Rorschach ink blots are used by psychiatrists and psychologists to learn about an individual's mind.

Rorschach ink blot: What does this look like to you? (Looks like a scary wolf to me, and like two ballerinas to a friend.)

The human is never just an observer of a scene, but an active participant in creating his perception. The viewer picks out what information is deemed important and what is not. When someone labels a photo as "a group of kittens," the label has grouped kittens together and disregarded other visible information (background pillow, wall, grass, toy). The picking out of kittens alone as the label shows us the viewer's priorities.

* * * *

A human does not and cannot simultaneously focus on all information in a scene. Humans don't have the mental capacity. Humans focus on some things and ignore others. When you enter a room, your eyes are drawn to something or things. Perhaps you focus on the gracious hosts, perhaps a statue to the side. If there is a rat in the middle of the floor, your immediate perception will be of the rat and not of the rose wallpaper.

If you enter the room and there is an attractive nude, you likely won't notice what is on the coffee table. You might not even notice the coffee table. After blushingly excusing yourself and scooting out of the room, you may not recall the existence of a coffee table, but it was there right in front of your eyes.

This focus, and the resulting perception, is your creation.

* * * *

Is it three bars or a horse shoe?

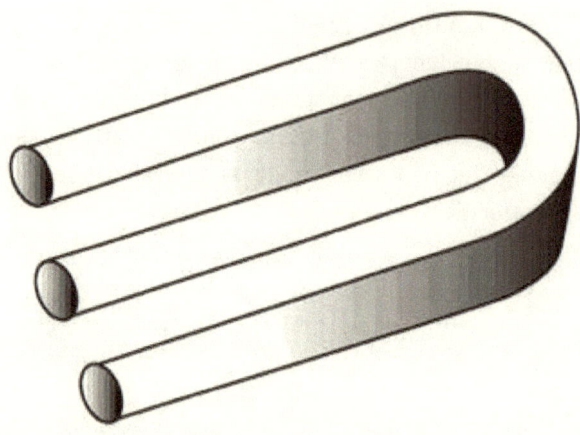

With the just shown *impossible trident* visual illusion, the viewer forms a perception about the whole from looking at just one end.   When she looks at the other end, she realizes her extrapolation was wrong.   Unlike the examples from the previous chapter (*Visual Illusions: Imagination*), there is no missing information.   All of the information is there for the eyes to see, but the viewer forms her initial perception as if information is hidden.   She mentally hides the information herself.

* * * *

If you ask someone to group and prioritize things in a picture, you will see both her biases and how she perceives. The perception will be just as much about the viewer as the picture.

* * * *

The viewer's purpose shapes perception. A person going to look at the art will have a different perception of the museum than someone merely stopping by to use bathroom. A kid visiting to do a report on sculpture will have a different perception of the art than a kid doing a report on paintings. If they visit different areas of the museum and enter different doors, they may have different ideas about the building's architecture.

The purpose is formed before the scene is viewed, meaning a perception is partially predetermined.

* * * *

Language
Language is a common way to organize, label and perceive objects and ideas. Native language is something we learned as infants, talk, think and even dream in. Our native language has profound influence on how we look at the world. Different languages give different emphasis, meaning, aesthetics, sounds and, perhaps most important, categories to things. As one perceives and thinks in part through categorizing (cats belong as one group, dogs belong as one group, magazines as another), native linguistic categories influence even nonconscious perception. It influences how we imagine things when our eyes are closed.

An elemental example of difference between languages is when a person in Atlanta Georgia and a person in Rome Italy read the same word 'pizza,' yet imagine different things. A pizza in Georgia is different than a pizza in Italy. If you asked the two to identify a pizza at a market, they might point to different objects. The Italian may say of the Georgian's choice, "You're crazy. That's not pizza. Let me read the label ... *Tombstone* ... *Do not defrost before cooking* ... *remove cellophane* ... *Glenview Illinois* ... You Americans might know Slim Whitman and Gilligan's Island,

but you know nothing about pizza. Come to Rome and I'll
show you pizza."

Many differences are more subtle. For example,
different cultures do not always categorize color alike.
Different languages can and do have a different number of
names for colors. This means a particular name, say red or
green, will apply to a different range of wavelength on the
visible light spectrum. It's the same total light spectrum of
color for both cultures, but the different numbers of names
divide the spectrum into a different size pieces. Like cutting
two identical pizzas, one into nine pieces and the other into
seven. The pizzas are identical except one has fewer and
bigger pieces.   In one culture, 'red' can cover a different
range of color than the equivalent word 'red' in another
culture. What you call red, a person on another continent
may or may not call red.

Even within a culture, people often categorize colors
differently. This is commonly done in the marginal areas,
such as aqua blue, dark orange versus red, magenta versus
pink. It is probable that you perceive some borderline colors
differently than your spouse, friend or co-workers. If two
friends define colors differently, they may believe they are
talking about different cloth swatches when they are talking
about the same one. Or they may believe they are talking
about the same swatch when they are talking about different.

This between friends difference can be because they
don't have the exact same color vision and that they never
had a serious discussion about what are the boundaries of
aqua blue, or what constitutes badious, brunneous and
gamboges. I don't recall ever having an instructor teach the
exact boundaries of aqua blue, aqua marine or magenta, not
even in art class. I doubt I ever had an instructor who knew
the exact boundaries.

As humans commonly communicate, learn and conceptualize the abstract through words, different interpretations of words often lead to conflicts. What may at first appear to be a visual illusion or even mental illness in a person may be a difference in culture.

An American joke is "Never ask for Squirt on an English airline." To Americans, Squirt is a brand of lemon/lime soda pop. To the English the word means urine.

I think it's safe to order 7Up.

So, if a tree falls when no one's around does it make a sound or doesn't it?

Many arguments are not caused by disagreement over the main ideas, but that the arguers unknowingly define terms differently. Arguers may have different definitions of war, peace, work week, formal attire, animal, automobile, tall, stiff drink and sexy, even though they both assume they are using identical definitions. Once the parties mutually set the definitions (which they didn't do in the beginning), they are often surprised to discover how much they agree with each other. Many arguments, many conundrums, many philosophical debates exist simply because parties never thought to mutually define terms.

An age old question is "If a tree falls in the woods and no one is around to hear, does it make a sound?"

The answer to this question depends on what is the definition of *sound*, and a key to the discussion is the determination of what sound means.

Is sound defined by the act of a human or other animal hearing? Or can a sound exist with none around to hear it? It would seem the smart thing to start by looking up the word *sound* in a dictionary.

I looked in one dictionary and two encyclopedias. One encyclopedia said that sound is defined by the ear detecting (hearing) the vibrations in the air. This would mean the tree in the question would make no sound if no one is around. The other encyclopedia and the dictionary defined sound as the vibrations itself, whether or not someone is around to hear them. By this definition, the tree would make a sound even if no one was around.

As you see, the famous tree debate isn't a matter of philosophy but of word definition. The difference between "Yes, it makes a sound" and "No, it doesn't make a sound" can come down to the arbitrary choice of definition, the outvoting of 2 reference books to 1, the flipping of a coin. Depending on what definitions used, the answer of Yes and No can describe the same forest scene. Is one sound definition superior than the other?   Not that I can see. They're just different.

People also have differing definitions of the word *one* in '…no one is around to hear…'   Some people think deer, birds and mice count as ones, while others think only humans count.   The definition of *one* can also determine whether the answer is question is Yes or No.

Certain words have strong connotations in a culture, and people intentionally play around with the definitions so they can apply words as they desire. If *patriot* is a popular label, people will fiddle with the definition so that they are defined as patriots and their enemies are not.   If patriot is an unpopular label, the same people would define the word so that their enemies are patriots and they are not.   These shameless self serving manipulations of definitions are common during political campaign season, but also during our daily lives.   What may be a *lie* when someone else does it, is a *fib* if you do it.

Notice these instances involve people being emotionally attached to a word no matter how it is defined.   It's word numerology.

When I was in high school, the quarterback for the football team came to school wearing a pink sweater. He spent the day saying, "No, it's coral."

(24)
Reactions Versus Answers

Definition of terms for this chapter:
Answer: The correct answer to a question.
Response: A reaction to a question that it is not an
    answer.

* * * *

To a question, there are two types of reactions: An answer and a response.

An answer is the correct answer to the question. A response is not the correct answer. While perhaps relevant to the question and offering useful information, a response does not answer the question. It could be said that to a question there is either an answer or anything else (a response).

Question: "What does 1 + 1 equal?"
Answer : "2"
Response : "I'm sorry, I don't know. I was never good at math. Give me a geography question."

The "2" is the answer. 1 + 1 = 2. The "I'm sorry, I don't know ...." is a response. It does not give the answer ("2") or attempt to give the correct answer.

Saying "Are you trying to insult my intelligence?" is a response to the question, rather than an answer. Saying "I don't have to answer your stupid questions" is a response.

Question: "Johnny, did you take a cookie from the cookie jar?"   (Johnny took a cookie from the cookie jar.)
Answer: "Yes, I did."
Response: "I did a lot of things today. I don't recall taking a cookie, but it's possible I might have taken it and forgotten about it. What kind did you bake?"
Response: "No."
Response: "What if I did?"

As shown above, while a response doesn't give an answer, it can offer information and even insight into the psychology of the responder. The response "What if I did?" neither answers nor attempts to answer the question, but reveals defiance.

* * * *

Many questions cannot be answered by humans. These questions usually are unanswerable because the answers are beyond human knowledge and sometimes comprehension.

"What is the exact number of grains of sand make up the Sahara Desert at this moment?"
"In square inches, what is the exact volume of the universe?"

* * * *

A leap of faith is a response.

* * * *

Many questions are unanswerable because the questions are worded so they are unanswerable.

Question: "What is the best color?"
Response: "I can't say what is the best color, but green is my favorite."
Response: "I don't know, but blue is probably the most popular."
Response: "Red."

There is no absolute, objective answer to what is the good, bad or best color. Any pick is subjective. Any definition of best is personal opinion. The first two responses offer perhaps useful information, but don't attempt to answer the question. The "Red" response also is a response but cloaked as an answer.

As with earlier unanswerable questions, the responses can give related information and reflect upon the question. The first two offer comments on colors. The matter of fact answer of "Red" shows arrogance or ignorance.

* * * *

The following are common unanswerable human questions:

"What is the meaning of the universe?"
"Why am I here?"
"What is my purpose on earth?"

There can only be responses to these questions. Religions and many political and social systems are responses to these and other unanswerable questions. They may present their responses as answers, but they are responses. Calling a response an answer is part of the response.

* * * *

Responses to unanswerable questions shouldn't be judged as answers, but as responses. *Considering it is impossible to know the answer, is this response to the question legitimate and reasonable? Is this response a fair way to respond to the unanswerable?*

I would classify the earlier response "I can't say what is the best color, but green is my favorite" as fair. It's not an answer, but a fair enough response to an unanswerable question. As suggested before, I don't think much of the "Red" as it's posing as the answer when there is none. The green guy is happy to give you his opinion, but readily acknowledges he doesn't have the answer. That seems to be a fair response.

I'm an art historian, and in art and collectible authentication, perhaps the number one rule is the expert should never make up an answer when he doesn't have one. He shouldn't say he's 100% sure, when he's only 75% sure. If you don't know, you don't know and, considering no one knows everything, there's nothing deficient about an expert saying he doesn't know. Find a self-proclaimed expert who has all the answers and you've found someone whose opinion you should be wary of. This should help explain why I didn't much of the "The best color is Red" answer. If she said "I have no clue" she would have gotten high marks. If she said "I don't answer dumb questions," she might have gotten even higher.

(25)
Mirages

Commonly associated with nature, mirages are visual illusions where what we see is correct, but abnormal. Mirages in nature are most commonly caused by unusual bending of light under unusual air conditions. The view can be so abnormal that the viewer 'can't believe his eyes.'

The most famous mirage is when it erroneously appears as if a pool of water is in the desert. More than a few thirsty wanderers have found nothing but disappointment ahead. The above pictured *water in the road* is the same type of mirage. Another related mirage is when sailors see an upside down ship in the sky. Enough to convince a pirate to swear off the hooch

These three particular mirages happen when there are abnormal layers of hot versus cold air that cause the light to refract, or bend, from its usual course. This bending causes an object to appear in an unexpected place. In the desert and highway a piece of the blue sky appears below the horizon, and is wrongly interpreted to be water. At sea a ship is bent upwards so it appears to be in the sky air.

* * * *

A mirage is called a **superior mirage** where the object appears above where it normally appears (boat in sky). An **inferior mirage** is when the object appears below the where it normally appears (sky in desert).

The inferior mirage happens when there is hot air near the ground. It shouldn't surprise that inferior images commonly happen when the ground surface is hot (desert, summer highway).

A superior mirage happens when there is cold air near the surface. They commonly appear in the arctic and over frozen water.

* * * *

Sunrise mirage. One of the most striking superior mirages is a sunrise mirage. These are seen over frigid areas, such as frozen lakes and seas. The light of the sun is bent upwards along the earth's curved surface making the sunrise appear earlier than normal. The sun is also distorted. Sometimes two suns are seen at once, one superimposed over the other.

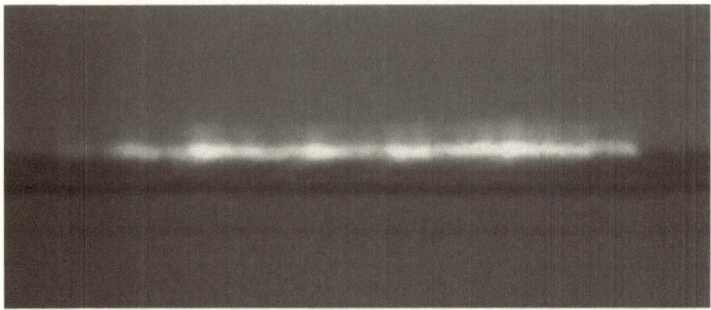

sunrise mirage seen from winter Minnesota

This mirage was noticed centuries ago by Western explorers stranded in the arctic over the winter. That far north there is no sun 24 hours a day for much of the winter. The explorers were surprised when the first sunrise of the season appeared days before it was supposed to. It wasn't until centuries later that experts realized the explorers had witnessed this mirage.

* * * *

Water can bend light just as air can, the light bending from air to water or water to air (or air to water to air, etc). A hardboiled egg distorts from normal appearance in a glass of water. The experienced spear fisher knows to spear to the side of the image of the fish or he will miss. Stones appear to ripple and wave in a crystal clear brook. One can study and demonstrate how mirages work with a drinking glass.

* * * *

The mirages aren't wrong views of an object, just different. Our normal vision involves distortions, including to color, details and angles, so one can hardly claim our normal vision is perfect and anything different imperfect. When they wish a better look, people with 20/20 vision intentionally distort their vision with magnifying glasses, binoculars, periscopes, video cameras and sunglasses.

When you view a bird through binoculars the lens distorts the light to make the bird appear larger and more detailed. You don't consider the binocular view of the bird wrong. You consider it to be more reliable than your naked eye view ("I thought it was a hawk, but it's just a crow.") A submarine's periscope bends light via mirrors so a sailor can see above water. The sailor doesn't consider the view make believe. He considers it a view of reality.

Humans classify views as mirages when they are abnormal and mysterious (at least to the viewer). There are many brilliant atmospheric effects that aren't considered mirages, as they are well understood. Little is more magnificent than a rainbow, but they are frequent and people know there is a scientific explanation. Fog, snow, sunsets and seeing our reflection in puddles would be considered astounding if they weren't common events.

That thousands of pounds of bright white snow changed into grass in one (hot) weekend doesn't cause you to write to Ripley's Believe It Or Not. You are well aware heat melts snow and underneath the snow is grass. You mowed that grass a few months ago. Ripley himself likely had this occur on his lawn numerous times. The changing of the season is impressive, but only a mirage to folks who have no memory of it.

After waking up in the morning and seeing the season's first blanket of snow, my very young sister turned to my dad and said, "Daddy, how'd you do that?"

When people move to new geographies they often experience new weather phenomena. When I moved to Seattle, I experienced unusual (to me) night lighting effects caused by Puget Sound and clouds. One night I thought there was a large fire on the other side of the sound. I later found out it was the lights of a distant hill-hidden town reflecting off of low clouds. This created a low, fiery glow. I see this lighting and it no longer fazes me. The first time I saw it, it was a mirage. Now it's town light reflecting off of low clouds.

You can't trust water: Even a straight stick turns
crooked in it.

    -- W.C. Fields

Heisenberg Uncertainty Principle

The Heisenberg uncertainty principle is a famous quantum mechanics principle made by German physicist Werner Heisenberg. Written mathematically, the word translation is that it is impossible to determine both the position and the momentum of a subatomic particle (an atom's electron, proton or neutron). The more accurately you measure one quality, the less accurate becomes the measurement of the other quality.

This happens because to measure the position of a subatomic particle you must shine light on it. The scientist needs to shine light to 'see' the particle, just as you or I need light to locate an apple or chair. While necessary to identify the position, the added light energy speeds up the tiny particle. The act of observing the particle changes it.

(26)
Visual Illusions: The Illusion of Depth in Two Dimensional Art

This 1400s Raphael painting uses many techniques to give the sense of   depth, including diagonal lines, diminishing scale, placing objects top to bottom.

Creating the perception of depth in paintings, sketches and photographs is a challenge, one that cannot be completely solved. This is because depth is three dimensional, while a sketch, photographic print or painting is two dimensional. Three dimensions cannot physically exist in two dimensions— they are mutually exclusive.

If you hold a crystal clear family snapshot of the Grand Canyon in your hand, at least logically you know that distant cliff and cloud is not miles behind your hand. You know it is just an image on the surface of a flat piece of paper.

Over the centuries artists have developed techniques to create the superficial representation of depth in 2D art. Before these techniques, paintings and sketches lacked any sense of depth. Cave drawings appear primitive as the artists didn't understand the standard concepts of depicting depth. An early European painting shows objects in unreal proportions to each other. A mile away person may be the same size as a person up close. People today would compare the proportions to 'kid's drawings.'

This chapter looks at a number of standard techniques used to give paintings, sketches and other 2D art the illusion of depth. These are techniques you can observe in art at the museum and incorporate into your own art. These are also 'techniques' you can observe in a real life, such as when looking at your living room or across your back yard. After all, the art is attempting to duplicate natural scenes like these.

* * * *

Overlapping objects

An object appears to be in front of the object(s) it overlaps. Overlapping is the strongest indicator of relative distance, overriding all other signs when there is seeming conflict. In the above Cezanne painting, the large center tree overlaps the *distant* bridge, mountain and sky.

* * * *

Diminishing scale
With things that are believed to be of same of similar size (2 cats or 2 basketballs), the visually larger appears to be closer than the smaller. In the Cezanne painting, the viewer assumes that the tree is much smaller than the distant hills. Thus the difference in scale (tree taking up more painting space than the hills) makes it appear as if the tree is closer. In the earlier Raphael painting, the smaller people appear to be further away than the larger. This is because the viewer is under the assumption people are of similar size when standing side by side. Remember the Adelbert Ames Jr card trick, where viewers incorrectly thought the physically larger cards were closer than the smaller ones? Ames' trick played on this diminishing scale bias.

* * * *

Diagonal lines representing diminishing scale
An exemplification of diminishing scale, diagonal lines moving towards each other as they move up or down a painting or sketch give the illusion of depth. A real world example of this is a straight road that appears to become skinnier as it approaches the distant horizon. Another example is when you stand at one end of an empty hallway and watch the lines where the wall and floor meet visually

move towards each other as they move to the farther side of
the room.

This photo shows diagonal lines and diminishing scale

* * * *

Colors

Without contradicting signs of depth, humans tend to
perceive bright, warm colors like red, orange and yellow as
being close, and dark, cool colors like blue and dark purple
as being further away. This is particularly true for abstract
images where there often is a lack of other depth or identity
clues.

For landscapes, adding blue will make hills and
mountains look more distant. The further away the bluer.
This mimics the real world, where distant mountains have a
bluish tone.

Bottom to Top Placement of Ground and Top to Bottom Placement of Ceilings

Barring conflicting information, humans generally perceive
what is at the bottom of painting to be in front, and what is at
the top to be in the back.   This is particularly true when
looking outside where there is no 'ceiling.'

Top to bottom: The bottom fans appear to be closer than fans and lights near the top. This is also an example of diminishing scale, with the bottom fans being larger than the top fans and lights

Inside a building, the ceiling can have the opposite effect, with the ceiling area nearest you appearing higher than the ceiling area further away.

In this room, the floor appears to move up the further it gets away from you. The ceiling (which is sort of like an upside down floor) appears to move down. These are both the product of diminishing scale.

* * * *

Focus

Things that are in focus tend to be perceived as closer than things that are out of focus. This makes sense, as a road sign is blurry if too far away.

Similarly, objects that have more intense color, detail and contrast often appear closer than objects that are blurrier, hazier and less focused.

In this old photograph depth is shown by diminishing scale, the narrowing lines of the road and building tops, and that with distance things become blurrier and hazier.

* * * *

Many visual illusions manipulate these techniques. The illusions often use incongruous, seemingly illogical techniques to toy with our minds. One quality suggests one thing, while another suggests the opposite. One quality evokes closeness, while another evokes great distance in the same object. The discord produces an emotional reaction in the viewer. The illusion will appear impossible to the viewer, and can literally raise his blood pressure and heart rate.

The natural signs of depth can also fool us in the real world. Nature can give seemingly conflicting signs. Houses appear larger and further away in heavy fog. In a movie, what appears as a full sized house or ship or dinosaur can be a miniature model. Carefully crafted sets make the things appear many times larger than they are. The moon appears larger when visually closer to the horizon. Rooms can be colored to appear roomier.

* * * *

A problem in trying to create realistic depth in two dimensions is that the human is designed to detect real depth not a flat representation. Looking at the real back yard, each eye looks at the 3D objects from a different angle, the head and body movement creating even more perspectives. The mind combines these different views into the mind's image.

This cannot be done with a two dimensional object. With a still life painting, and even a still life photograph, it is not possible for the eyes to get the different views of the fruit bowl that is needed to perceive a truly 3D fruit bowl. The photograph, no matter how clear, shows only one angle.

Notice that many attempts to create a closer to true 3D effect involve alteration not just to the flat image but of the viewer's vision. 3D movies and pictures often require special glasses and viewers.

The hologram is a rare example of a flat image that can realistically simulate three dimensions, allowing the viewer to see angles and even sides of the pictured object.

* * * *

Cubist paintings, where different sides of an object are seen simultaneously, can be looked at as an attempt to represent 3D in a 2D plane. A cubist painting sometimes also represented the passage of time, with a person being shown at different times.

Visual illusions point out the existence of blind spots and unreliability in our logic and reasoning systems.

For those who have never before seen this image, the rational answer would be the bar changes in tone. To say it is solid in tone would be irrational.

It's not that all false perceptions of reality are due to faulty logic, but that many are formed using what is considered sound logic and reasoning.

Humans categorize and label objects in part by visible colors. Many animals, flowers, gems and even humans are defined by their visible colors.

As defined by the American Kennel Club, a cairn terrier can come in all colors except white. If a cairn terrier is born white, it's not a cairn terrier. It's a West Highland Terrier, a different breed.

If we could see infrared and ultraviolet light our categorizations and objects, including terriers, would be different.

(27)
Illusions That Can't Be Escaped

The two balls are the same size

Even after you learn how they work, there are many visual illusions that still fool you. If you go back and look again at the visual illusions shown throughout this book, many will still fool your eyes. The two identical balls in the above picture appear to be different sizes even though they are not. If you look at the picture tomorrow or two months from now, they will still appear to be different in size.

The mind contains compartments that perform specific tasks. For example, one compartment is used for comprehending spoken language, another for perceiving smell. Some of these compartments are isolated from other parts of the brain. They work on their own, not influenced by goings on elsewhere. These compartments sometimes are even isolated from conscious knowledge.

The perception of many visual illusions is made independent of your conscious knowledge. This explains why even your conscious knowledge that they are illusions doesn't solve your nonconscious misperception.

Even though you have already learned it is of only one color and tone, the above horizontal bar still appears to change. The visual perception is made independent to your conscious mind.

Simplicity

To humans, simplicity is that which is simple to them. Simple matches one's sensibilities, knowledge, intuition and expectations. If it didn't, it wouldn't be simple. What may be simple to one human may not be to another. What may be simple to humans may be simple only to humans.

Simplicity has long been used by humans to define supposedly absolute things like cosmic truth, goodness, beauty, logic and purity. There are a number of problems with this. One is there is no proof that cosmic truths, for example, are simple. Another problem is simplicity, and thus what is defined as cosmic truth, is in the eye of the beholder.

Normal, even nonconscious thinking involves simplification, translating complex information into something understandable. Conceits are simplifications.

Your visual perception involves simplification-- interpreting a complex scene, grouping and labeling the objects according to your experience, focusing on what you seem to recognize and ignoring what you don't. Visual illusions and mirages shown throughout this book involve simplification. The scene or graphic is translated by the viewer into something understandable, an understandable translation that happens to be wrong. This alone proves that simplicity is not proof of truth, and that truth isn't always simple. Lies are often simpler than truths.

Simplicity, of course, has many practical uses. Scientists strive for simplicity in theories and testing. A scientific theory that is needlessly complicated will needlessly confuse students and seasoned scientists alike. Needlessly muddled theories are harder to test, study, correct and understand. In our daily life, good verbal communication requires simplicity, including using words, phrases and language the listener understands. If a traveler speaks only English, it does them no good for you to give road directions in Spanish. Road directions in Spanish may be simple to a Spanish speaker, but it's complicated to someone who doesn't know the language.

(28)
Values, Culture and
Aesthetics in Visual Perception

Give an objective identification of what is in the following three pictures. Answer one picture at a time, by saying the answer aloud or to yourself. <u>The images are not digital tricks or manipulations</u>. They were picked because of their straight forward, familiar subjects.   I am just looking for quick objective identifications.

One or more of your answers likely was on the order of 'George Washington crossing the Delaware,' 'a bald eagle' and/or 'a watch.' These answers are not objective, being formed in part by value judgments, aesthetic views and other personal biases.

In the lower left picture there is much more than a bald eagle. There is sky, stump, trees. The 'eagle' answer subjectively singles out one thing. Part of this is due to a personal and cultural value judgment that a bald eagle is more important than the other objects. Another reason is because the eagle is pictured large, clear and centered. If the picture showed a tree close up and in focus and a small out of focus eagle flying in the distant background, your answer likely would have differed. Change in arrangement, size and focus effects the viewer's labeling, even when the identical objects are pictured.

You may not have known the dark blurriness near the bottom is trees, but that does not change their identity. If you called them bushes, that would not make them bushes. It's common to ignore the unknown.

Similarly, if your answer to the lower right picture was 'a watch,' you made an aesthetic and value judgment about what is and is not important. Placement and focus affected your judgment, along with your feeling that a potentially expensive watch is the center of attention.

In the top image there are quite a few people pictured. If you answered "George Washington crossing the Delaware" you singled out one as being the identity. This is in part due to a higher value placed on George Washington, a famous figure in United States history. This is also due to your knowledge, as Washington is likely the only person you know by name. Again, it is common to focus on the known and ignore the unknown.

If you said "This pictures a bunch of people, one whose name is George Washington" you would have given a broader answer, while acknowledging the extent of your knowledge.

Also notice that your answer was not 'sky, water and ice,'

even though sky, water and ice takes up more space than the men, boat and flag. This was due to your bias that the human is the natural center of attention.

The initial request of this chapter was to give objective identifications, but your answers were subjective. I didn't ask for your moral judgment of George Washington versus other men, whether a bald eagle is more significant than out of focus background trees or the relative financial value of a watch.

* * * *

These and other types of subjective judgments are both natural and essential to humans. Quick interpretations of scenes, including judging what is and is not important, is essential to getting through our day to day lives. You wouldn't have lasted long on this earth if you placed equal visual significance on a twig on the pavement and a car speeding in your path. If someone unexpectedly tosses you a ball, you catch the ball by focusing on it. If you focus on the thrower's shoes or what's on TV, it is probable you will drop the ball.

The problem is that, while essential, this type of subjective identification helps make it impossible to make objective identification. One's identification is always shaped by one's knowledge level, past experience, aesthetic view, pattern biases and value judgments. As shown with the identification of the three pictures, the human is often not aware of this influence. To many people, biases are what others have.

Measuring the reliability of the mind

To us humans, the reliability of the human mind cannot be known, as we use the human mind to test and judge the reliability. If your goal is to determine the accuracy of the human mind, that means you do not know the accuracy of the tool used for testing and judging (the human mind), which makes it impossible to determine the accuracy of the human mind.

Your opinion about the reliability of the human mind involves a leap of faith. A common tendency is to overestimate the reliability. There are a number of reasons for this. One is that many errors and blind spots in thinking are unknown and not counted. Another is that a human's belief system and world view are premised on a reliable mind. If the mind's reliability comes into question, so does the reliability of the belief system and world view.

29) Logic Versus Art, Facts Versus Fiction in Expressing Higher Ideas

Two lovers were cursed, he to be a wolf at night and she to be a hawk during day. They could not be human together.

Humans perceive, interpret and mentally explore their world on many levels. Humans experience things rationally, irrationally, consciously, nonconsciously, emotionally, intuitively, directly, indirectly, aesthetically, figuratively, literally-- in a varying combination of these and more all at once. A human can think logically one moment and be emotionally swept up by a song on the radio the next. Math professors fall head over heels in love and abstract painters calculate their taxes. A great movie is sometimes enjoyed on the intellectual and emotional levels.

A human's best possible exploration, understanding and expression of the universe uses all the levels. An expression of the universe through only mathematics or only music is inherently limited. Many things in the world can't be explained with mathematics-- love and beauty for examples--, just as mathematics can't be explained with love and beauty. An explanation using just one level is flawed.

This chapter touches on two standard and distinct methods for making profound explorations and representations of the complex world: the logical essay and art. One is based on reason (logical essay). The other has its meaning in the irrational (art). Each is a legitimate method of communication yet limited in what it can express.

* * * *

Logical essay

Humans often find it important to explore subjects and ideas logically. A logical essay uses reason and, well, logic. It intentionally tries to remove emotion, whims, logical fallacies, subjectivity and, except when clearly identified as such, the author's opinion. The language itself is expected to be free of logical fallacies and linguistic muddiness.

* * * *

Logical essay

In proofing the logical essay, the writer and proof reader make sure that statements are consistent. As statements are built upon statements, even small logical fallacies can undercut the entire essay.

The following are examples of checking the logic of statements.

Statement #1: "Jenny has only one brother. Thus, her brother has only one sibling."

Analysis of statement #1: Incorrect, should be rewritten. If Jenny has a sister, then the second statement would be untrue, as John would then have more than one sibling. While John may indeed only have one sibling, the first sentence does not prove the second.

Statement #2: "Jenny's favorite type of fruit over all other fruit is the orange. Thus, the banana is not her favorite fruit."

Analysis of #2: The statement is logically correct.

* * * *

Art

Opposed to the logical essay, the essential meaning of art is based in irrationality. While a work of art has an underlying and often even logic-related structure, the essential meaning is irrational (sublimeness, profound beauty, aesthetics, emotional response). Art produces a profound psychological, sometimes visceral effect on the audience and it is here where the meaning exists.

This irrational meaning is illustrated by the wordless music you love. There is nothing logical or rational in the sounds or the emotional reaction you get from them. Art's meaning exists beyond logic and reason.

* * * *

Art

Artists intentionally subvert logic, reason, objectivity and reality to produce the desired psychological effect in the audience.

Many paintings intentionally distort reality. Look at paintings by Picasso, Dali, Cezanne, Jackson Pollock and Renoir. Even the 'realistic' paintings of the 1300s have impossible dimensions, odd looking humans and made up visual stories.

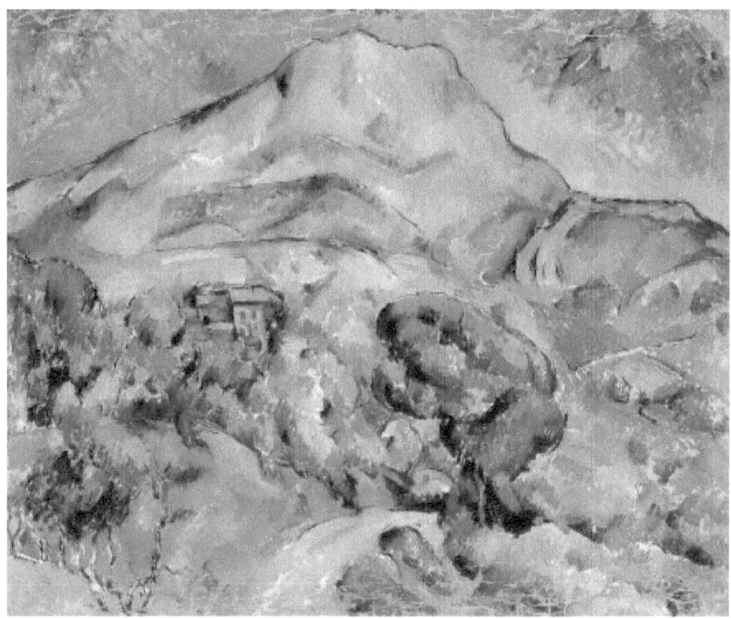

Painting of nature by Paul Cezanne

Classic movies and novels have unreal plots, characters, timing and effects. Some are fairy tales and some are science fiction.

To produce the desired emotions in the audience most movies have music sound tracks. In real life many of the scenes portrayed would have no full symphonic accompaniment. Washington crossing the Delaware, man lost alone in the middle of the desert, Humphrey Bogart walking a deserted street, bear fight, Rocky Balboa running up the Philadelphia steps. Most movie music is an intentional distortion of reality for psychological purposes. Isn't there something bizarre about musical accompaniment for a National Geographic documentary about insects? What does synthesizer or orchestral arrangement have to do with ants?

Even a literal-minded scientist will complain that a documentary about physicist Werner Heisenberg didn't have a

music soundtrack, or that the music wasn't what he would have chosen. If asked, he might tell the director he would have preferred Beethoven over the used Bach, perhaps mixed with some Mozart.

* * * *

Art is so different than the real world that its truth is derived from lies. Shakespeare's Hamlet is made up. Of Mice and Men is a figment of John Steinbeck's imagination.

* * * *

An irreconcilable conflict exists between art and logic. One requires rationality and the other requires irrationality. Each subverts the other.

* * * *

Logic

An inherent problem with the logical essay is that, despite the author's intentions, it can never be free of the things the author wishes it to be free of-- subjectiveness, irrationalness and arbitrariness.

The author has subjective views about writing style, structure, pacing, aesthetics and overall presentation. A writer can't write or think without using a plethora of conceits, some chosen, some nonconscious, some inborn. A writer can't visualize things in his mind without biases, personal and cultural ways of grouping, labeling and conceptualizing. Writers take into consideration the conceits of the audience, as the point of the essay is to communicate.

Even the seemingly perfectly logical equation $1 + 1 = 2$ demonstrates human taste in the even spacing, balance, linearity, colors. Many would rewrite "$1+1 = 2$" as "1 + 1 = 2." The equations mean the same thing, so the reason for the change is aesthetic. Pure mathematicians will be the first to tell you that math can be beautiful and ugly, and that their research is influenced by aesthetics.

In practice, human logic has its own art.

* * * *

Fact versus art in the biography
The subject of the biographical movie or book is or was flesh and blood, a life filled with measurable facts: dates, times, durations, amounts, heights, geography, quotes, test scores, employment records, hair color, mailing addresses. Yet a strict recitation of facts will not wholly represent the person and her life (much less entertain the audience). A person is much more than facts and dates. Character, personality, aesthetic vision (perhaps the subject was a great artist), beliefs, faiths, mental conflicts, contradictions, urges, dreams, fears, subjective experiences, nonconscious, desires.

A famous composer might say, "If you want to know who I am, listen to my music. That's all you need."

A woman might say, "If you want to know about me, forget about my high school transcript and the conversations I have with my boss. Watch my favorite movie. If you don't get the movie, you'll never understand me." Her favorite movie probably was made by someone she never met, perhaps who died before she was born, the movie isn't about her, perhaps takes place in a country or even planet she's never been too and may not have a single character that resembles or acts like her or even speaks her language.

Even when distorting facts and logic and time, a biography that is a work of art can be a better representation of the subject, his deeper personality and vision. This type of biography is a figurative representation of the person, as the earlier Cezanne painting is a representation of a landscape. Cezanne didn't intend or expect for the viewer to take the painting literally.

The essential problem in the biography is that to create this figurative truth, one must distort the factual truth. And to tell the factual truth, one destroys this figurative truth. The biographer needs the two to exist together, but they cannot.

30) Claire: How We Met

I met Dereb in college in one of those friend of a friend situations. He lived with some guys at a large apartment near the engineering school, and a friend of mine had gone to high school with one of them. We stopped by the apartment one afternoon because my friend had to pick up something. Some of the guys including Dereb were playing nerf basketball with a little plastic basketball hoop suction cupped to a banister. My friend's friend briefly introduced me to the guys. Dereb said hi or acknowledged me and my friend and went back to playing, and my friend and I left like a minute later. That was it. To be honest, I thought he was good looking, but didn't think much about it.

Not long after, like maybe two days after, I was standing in the snow outside the Union near the street. It was one of those bright sunny winter days. Dereb walked up to me and asked if we had met earlier, and I said yes.

31) Going Away

The medicine isn't intended to cure, but make me forget. Or, forgetting is considered the cure.

* * * *

The night has always been my world. It's not a matter of explaining how or why, it just is. Even when I was a little kid, my parents called me a night hawk.

Though identical to day on the map, night is a different world. A different fauna, flora, plot, music, smell, temperature, meaning. Daylight society is unconscious.

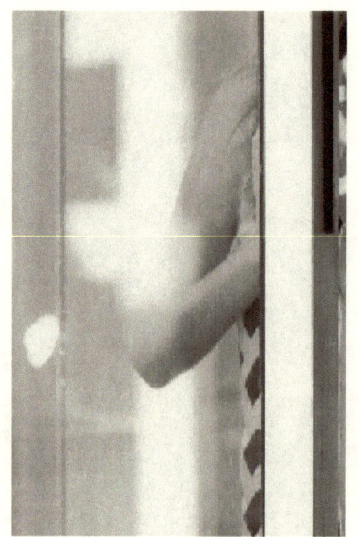

Later Pieces

(a)
Numeral Systems and Psychology

In some Western Hemisphere high rise buildings there are no thirteenth floors. Well, there *are* thirteenth floors, but the floors are labeled 10, 11, 12, 14, 15 to give the superficial appearance of having no thirteenth floors. The building owners know many have a superstition against the numeral thirteen and it's easier to rent an apartment or office if it's called 'fourteen.'

In Korea and Japan where four is considered unlucky as it's the sign of death, some buildings 'omit' the fourth floor.

* * * *

Our base-10 numeral system
The common modern human counting system— the one you and I use-- is based on ten, and is referred to as **base-10**. It uses 10 different numeral symbols (0,1,2,3,4,5,6,7,8,9) to represent all numbers, and many popular groupings are divisible by ten: 10, 20, 100, 300, 10,000, century, decade, top 10 lists, golden anniversary, etc.

Our base-10 system is based on the number of digits on a human's hands: eight fingers and two thumbs. As with today, many ancient humans found fingers and thumbs convenient for counting and it seemed only natural to base a counting system on the 10 digits.

While the base-10 is a good system and has served us well, ten as the base was a somewhat arbitrary choice. Our numeral system could have been based on 3, 8, 9, 11, 12, 20 or other number. Instead of basing it on the total digits on a pair of hands, it could have been based on the points of an oak leaf (9), the sides of a box (6), the fingers on a pair of hands (8). These different base systems would work. Some might work as well or better than our base-10 system. Nuclear physicists and tax accountants could make their calculations using a 9 or 11-base system. Once you got used to the new system, you could count toothpicks and apples just as accurately as you do now.

Quick comparison: counting with base-10 versus base-8

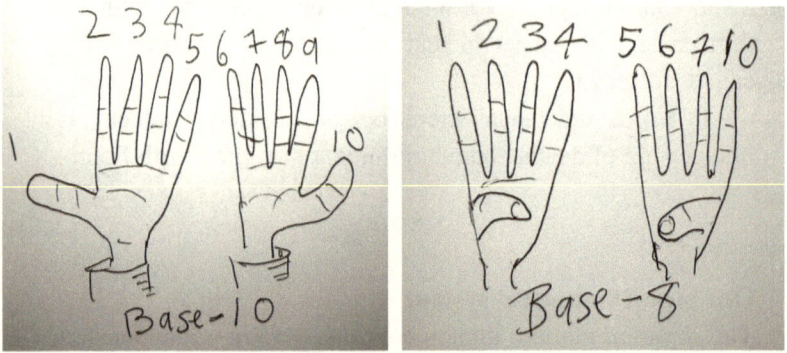

The above pictures compare counting with a base-10 system based on the ten digits of the hands (fingers + thumbs), and with a base-8 system based on just the eight fingers (thumbs not used). Notice that the base-8 system, not using the thumbs, is missing two numeral symbols: 8 and 9.

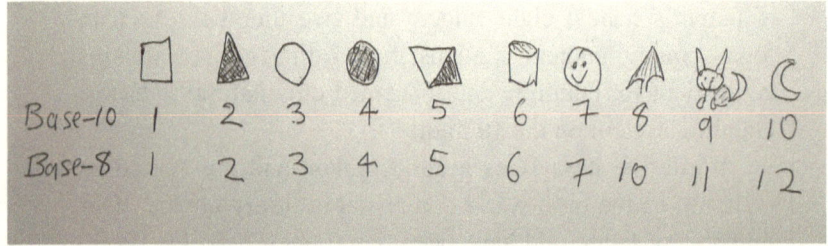

This comparison picture shows how assorted designs (top row) are counted with the base-10 and with the base-8 systems. As base-8 omits the two symbols 8 and 9, '10' comes sooner when counting in base-8. In one numeration system, the cat is '9' and in the other is '11.' As you can see, the real value of 10, amongst other numeral symbols, is not an absolute. It depends on what base is being used.

* * * *

Another example of counting with different bases

The below table illustrates how you can count symbols (far right column) using the base-10, base-9, base-8 and base-5 systems. If you wish, the symbols can represent physical objects like fruit or cars or plants. In this table the symbols are constant, while the numeral systems create different numeral labels for the symbols (or fruit or cars or plants). For those who consider '13' unlucky, notice that each counting system labels a different symbol as being 13.

Base 5	Base 8	Base 9	Base 10	Symbols
0	0	0	0	
1	1	1	1	!
2	2	2	2	@
3	3	3	3	#
4	4	4	4	$
10	5	5	5	%
11	6	6	6	^
12	7	7	7	&
13	10	8	8	*
14	11	10	9	(
20	12	11	10	)
21	**13**	12	11	-
22	14	**13**	12	+
23	15	14	**13**	"
24	16	15	14	:
30	17	16	15	>
31	18	17	16	<
32	19	18	17	{
33	20	19	18	]

* * * *

This counting stuff is not idle abstraction. Civilizations have used and use different numeral systems.

The Yuki Indians of California used a base-8 numeral system. Instead of basing their system on the digits on their hands, they based it on the spaces between the digits.

The Ancient Mayans used a base-20 system, as they counted with the digits on their hands and feet. They lived in a hot climate where people didn't wear closed toe shoes.

Today's computer scientists use 2, 8 and 16-base systems. For some mathematical work base-12 is more convenient than base-10. For this base-12 system they usually use the normal 0,1,2,3,4,5,6,7,8,9 numerals and add the letters a and b to make twelve (0,1,2,3,4,5,6,7,8,9,a,b). It goes without saying that these mathematicians, often university professors and researchers, are using this system to perform higher levels of calculations than you or I perform in our daily lives. They aren't counting change at the grocery store.

Our normal lives show the vestiges of ancient numeral systems. We sometimes count with Ancient Roman numerals (Super Bowl XXIV, King Richard III), letters (chapter 4a, chapter 4b, chapter 4c... Notice how this combines two different systems, standard numerals with letters) and tally marks. We group loaves of bread, inches and ounces by the dozen, and mark time in groups of sixty (60 seconds per minute, 60 minutes per hour). Counting inches and ounces by twelve comes from the Ancient Romans. Our organization of time in groups of 60 comes from the Sumerians, an ancient civilization that used a base-60 system.

The traditional counting of bread into groups of twelve has practical convenience. At the market, a dozen loaves can be divided into whole loaves by two, three or four. Ten loaves can only be divided by two into whole loaves. Sellers and customers prefer the grouping that gives more whole loaf options, not wanting a loaf to be torn apart. This should give you an idea why feet and yards are divisible by twelve, and there were twelve pence in a shilling— you get more 'whole' fractions out of twelve than you do ten.

These have been just some examples of other numeral systems, as there have been a wide and varied number over history. This not only includes systems with different bases, but with different kinds and numbers of numeral symbols. In Ancient Eastern countries, physical rods were used to represent numbers. The number, position, direction and color of the rod represented a number. In Ancient Egypt, pictures, known as hieroglyphics, were used to represent numbers. One thousand was written as a lily, and 10,000 as a tadpole. The Ancient Hebrews had a similar system to ours, except they used 27 different symbols to our ten. For the Hebrews, numbers 20, 30, 40, etc each got its own unique symbol.

Ancient Egyptian numerals for 1,000 (lily flower) and one million (man with raised arms)

Tallying is an ancient basic counting system many of us use. The practical problem with this system is that numbers like 500 and 10,000 require a whole lotta tally marks. 500 requires 500 tally marks. Over history, numeral systems have changed and evolved to correct inconveniences like this. Notice we use the tally system only for simple tasks, like keeping score in a ping pong game and marking days.

* * * *

A kid's counting system: Eeny meeny miny moe
Kids have long used counting rhymes to decide who is *it*. The below common rhyme does the equivalent of counting to twenty, with the last word being the twentieth word.

> Eeny, meeny, miny, moe
> Catch a tiger by the toe
> If he hollers let him go,
> Eeny, meeny, miny, moe

There are a few interesting things about this eeny meeny counting system. First, it is quasi base-20, not our normal base-10. Second, words are used as numerals, or as the practical equivalent of numerals. Kids could count to 20 for the same practical result, but they chose to use words. Third, while lucky 7, 10 and unlucky 13 have popular importance compared to other numerals in our base-10 system, the seventh, tenth and thirteenth words in the rhyme do not.

This is an example where a different counting system changes what numbers are perceived as important. Most kids who count with this rhyme aren't even aware which are the seventh, tenth and thirteenth words.

Humans often say they can't conceptualize numbers in anything but the normal base-10, but here is a base-20 words counting system that we have all used. Granted this counting system is simplistic in the extreme, used for one and only one purpose— to count to twenty (moe). You wouldn't want to try and use it to calculate your taxes.

* * * *

Numerals and human psychology

Humans form psychological attachments and biases for the numeration systems they use. Having grown up using a particular system, and seeing all those around them using the same, many people assume their numeration is absolute and eternal. Before reading this chapter, you may not have known or thought about the existence of other systems. Your base-10 system was all you knew, the prism which you saw the universe. 10, 100 and 1000— popular products of your base-10 system— are numbers you are attracted to. Thinking in base-8 or base-7 is foreign.

It's telling to look at how humans change their perception from system to system, and how a change of numeration system changes peoples' perceptions of things. The perception is not just about the numeration system itself, but the things the numeration system is used to count— objects, time, ideas.

* * * *

As the earlier tables showed, a different base numeral system doesn't change the accuracy of our calculations or the physical objects we calculate. However, if we retroactively changed our base-10 system to a non base-10 system (like say the Yuki's base-8 system) we would change how humans perceive and react to objects and concepts.

As with the high rise buildings and the superstitious renters, the historical changes would be caused in large part by human perceptions of the numerals themselves rather the things the numerals represent. No matter what the Mexico City building owner calls the thirteenth floor, it is the same floor. If he changes the label on the elevator directory from '13' to '9988' or to '789' or to 'Q,' it is the same floor with the same walls, ceiling and windows and distance above the sidewalk. The numerologist apartment seekers aren't reacting to the floor but to the symbol '13.' It should not surprise that a change to the symbols, such as caused by the changing to a new counting system, will change their reaction to the floors, along with many other things.

With a large lot of stones lined up on a table, changing the numeral system has no direct effect on the amount or physical nature of the stones. With a new counting system, the stones would be the same stones, but many to most would be assigned different numeral names. While the stones are the same stones no matter what we call them, human perceptions of the stones change as the stones' numeral names change. Under our popular base-10 system, humans consider certain numerals to be special, including 10, 100, 1000 and 13, and react accordingly to objects labeled with these names. With the new numeral representations, humans' perception and treatment of the stones will change. If before a person avoided a stone because it was unlucky 13, in the new system a different stone would be called 13. If in the old system the stone labeled '100' was singled out as special, in the new system '100' would represent a different stone.

If a human is asked to count and group the stones, the grouping will change with the different counting system. In the base-10 system, it's likely the person would make piles of 10 or 25 stones or similar standard. In an 8 or 9 base system, the number and size of the piles would be different. To someone standing across the room, the rock design would be different. Her aesthetic reaction to the formation would be different.

This shows that your numeration system isn't just an objective observation system, but helps form how you perceive objects. Under a different system, you would perceive things differently.

The lines separate the same number of coins. The left group contains 30 total coins in stacks, the middle group between the lines has 30 coins in stacks, the group to the right of the right line has 30 coins in stacks. The coins of each group were stacked by different numeral systems. This is why the same numbers of coins look different.

* * * *

Changing numeral systems, changing history
As a numeration system changes how we perceive, organize and react to things, a retroactive change to the numeral systems would change human history. The amount and type of change can be debated, but today's history books would

read different.

With a change to the standard numeration system, time would remain the same but human marking of time would change. The decade, century and millennium equivalents would be celebrated at different times. No Y2K excitement at the same time as we had. Special milestones, like current marriage 10th or 25th anniversaries, would be at different times. People who now receive 30 years of service awards might receive equivalent awards but after a different duration.

Think of all those sports championships decided in the last moments, including the improbable upsets and bloop endings. If the events took place at different times and under different numeral influenced conditions some of the outcomes would be different. If an Olympic sprint is decided by a fraction of a second, it's unlikely the first to last place order would be identical if it took place the day before with the runners in switched lanes and running a different length race. The changes to marking of time and distance would likely result in different gold, silver and bronze medal winners over the years. If a horse race was a tie, it is unlikely the same horses would tie if the race had been run earlier or later in the day or on a different day over a different length race. Realize that the change to the numeration system would likely change the standard race distances, even if the changes were just slight.

Think of all the razor close political elections. If the elections took place at a different time, even if just a day earlier or later, it's possible some would have different outcomes. A few of the outcomes could have been for President, Prime Minister, judge or other socially influencing position. Think of all those close historic battles that may or may not have had a different outcome if started at different times, using different size platoons and regiments and

Generals who made decisions using different number biases. Napoleon Bonaparte was superstitious of 13 and made his government, social and military plans accordingly. Think of the influential or not yet influential people who died at relatively young ages in accidents, from Albert Camus to General Patton to Buddy Holly. James Dean died in a sports car crash at age 25. Would he have crashed if he started his drive at an earlier or later time? Popular perception of the actor no doubt would be quite different if we watched him grow old and bald.

The powerful nineteenth century Irish Leader Charles Stewart Parnell would not sign a legislative bill that had thirteen clauses. A clause had to be added or subtracted before it could become law. Irish law would have been different under a different numeral system.

* * * *

United States consumer prices would likely be affected by a different numeral system, if just marginally. Again, this would be due to human psychological perceptions of numerals.

Even though most current US sellers and buyers think nothing of one penny, often tossing it in the garbage or on the sidewalk, sellers regularly price things at $9.99 instead of $10, and $19.99 instead of $20. Check the newspaper ads. This pricing is purely aesthetic, intending to play on consumers biases towards numerals.

The shallowness of this 1 cent game is illustrated when it is used by stores that have a 'give a penny, take a penny' tray, and that it is used in many states with different sales tax rates. Most people psychologically affected by $9.99 pricing at home are also affected by $9.99 pricing when traveling by car across the country. That the daily change in sale tax charge dwarfs the one cent between \$9.99 and \$10, illustrates the traveler's irrationalness.

Under a base-9 numeral system that omits the numeral '9,' \$9.99 and \$19.99 would no longer exist, and the visually appealing "one cent below big number" pricing would land elsewhere. In a 9 digit system, it's likely that there would be many \$8.88 and \$18.88 pricings in newspaper ads, and the same types of travelers would be attracted to \$8.88 and \$18.88 prices as they go state to state even though the taxes change state to state.

* * * *

There are a variety of intertwined reasons behind irrational biases towards numerals and numeral systems.

One reason is people form psychological attachments towards a system, its symbols and the standard groupings of objects made from the system. A three digit numeral price (\$9.99) looks distinctly different than a four digit numeral price (\$10.00), literally being shorter. One hundred stones grouped into 10 groups of 10 each will look different than 11 groups of 9 stones each with one left over. It's the same amount of stones, but their physical designs look different. There's an aesthetic aspect to how humans view symbols and groupings.

Closely related reasons are tradition and habit. If you have used our base-10 system all your life, it's as natural to you as your native spoken language. In fact words like nine, ten and decade are part of your daily vocabulary. If everyone you know uses this numeral system, the idea of using a different system may not have even crossed your mind before now. The idea of calculating using a base-8 or base-11 system seems strange and even unnatural to most people because they were raised on base-10.

Another reason behind irrational biases towards numerals is the seeming, if nonexistent, absoluteness of the familiar numerals. While the true nature of time, supernatural, war, love and the cosmos are shrouded in mystery, the numerals traditionally used in representing these things seem tangible, concrete. Unlike philosophical abstractions, numerals can be written down and typed into the calculator. Even little kids can count numerals on their fingers. That folks like Isaac Newton and Albert Einstein used these same numerals seem to numerologists to indicate the numerals' potency. Though, if asked, both scientists would agree they could have used other numeral systems to do their work, and there was nothing uniquely special about the system they adopted.

Numerals are used only as convenient notations, proverbial post-its to label objects. They have no absolute, inborn connection to the things they represent. Whether you call the animal cat or gato it's the same animal, and whether you call a number 5, five or V, it's the same number. Whether you count a grove of trees with a base-10 or a base-8 system, they are the same trees. If you count and label the trees a,b,c,d,e,f,g, they are still the same trees. Numerologists incorrectly assign an absolute meaning and identity to the numerals that doesn't exist.

Even in academia, mathematicians considered to be too enamored with the beauty of numbers at the expense of practical use are sometimes derogatorily called numerologists by applied scientists like engineers. Mathematicians are as influenced by aesthetics as the rest of us.

* * * *

Sounds Good

Many Chinese judge numbers as good or bad by what words they sound closest to. As their pronunciation of 3 sounds closest to their word for 'live,' 3 is considered good. Their pronunciation of 4 sounds close to their word for 'not,' so is often considered negative.

China is a huge country with many dialects. As numbers and words are pronounced differently in different areas, a number's perceived goodness and badness depends on where you are. For example, 6 is considered good in some places and bad in others.

(b)
Movement Illusions

Perception and misperception of movement is similar to the perception and misperception of still images. The viewer sees a limited amount of information from a scene and, using its experience, knowledge, biases, internal mental abilities and logic, makes a guess at what is going on. Often, this guess is correct, or at least a good approximation. Other times it is wrong.

Except for more extreme situations (very slow movement, very small objects), the human optical system is good at detecting the presence of movement. The misperceptions most commonly happen in the interpretation of the movement. Humans can correctly detect the presence of movement but misinterpret the direction, speed and even what is moving. A human can think object A is moving, when it is object B that is moving.

The following are two common examples of correctly

detecting the presence of movement, but misinterpreting it.

The parked car prank

A prank you have probably heard about is where two pranksters park their cars on each side of an open parking space. Sometime later the unsuspecting victim parks his car between these two cars. When the victim is fiddling with his keys or checking the contents of his wallet or looking in the glove compartment, the two pranksters suddenly drive forward in unison. The victim gets the instant sensation that his car is moving backwards and panics. He soon figures out what is going on, but is embarrassed. This is an example where a person correctly identifies movement, but misinterprets what is moving. Also note that his misperception was influenced by instinct, the victim having little control over the adrenaline rush.

Baseball's changeup

In baseball, pitchers use the so called changeup pitch to fool the batter. A changeup is intended to look like a fastball, but is slower. The changeup is typically thrown after a fastball, often after consecutive fastballs. Then, seeing the normal fastball arm and body motion of the pitcher, the batter believes the ball is again coming fast and swings accordingly. When the changeup works, the unexpected speed results in the hitter making feeble or no contact with the ball.

As pitching great Warren Spahn said, "Hitting is timing. Pitching is upsetting timing."

* * * *

Stroboscopic Movement Illusions

When watching an old Western movie there is a curious effect that sometimes stands out. The wheels of a moving wagon sometimes appear to be still, rotating slower than they should or even rotating backwards. This happens when the rotation speed of the spokes was not in synchronicity with the speed of the film.

The above three still images of a wagon wheel look to show the wheel in the same position, but they show the wheel at different rotations. The middle picture was rotated 90 degrees from the left image, and the right image is rotated an additional 90 degrees. That each spoke is shaped and colored identical to the others is an essential contribution to the illusion. If these were the stills in a movie the rotating wheel would appear to be motionless. If they were the stills in a movie, but the rotation was 80 degrees instead of 90, the wheels would appear to be going backwards.

The wagon wheel illusion in a movie is an example of the **stroboscopic effect**. In the dark, a strobe gives off intermittent flashes of light. Under a strobe, the viewer views a moving object though short intermittent snapshots instead of a continuous view. This can lead to misperception of the object's movement, as the viewer nonconsciously imagines what is going on in between the flashes.

Say you are watching a swinging pendulum under stroboscopic lighting. If the strobe flashes a quick burst of light once every second and it takes the pendulum exactly one second to swing back and forth, the pendulum will appear to you to be motionless. Each flash catches the pendulum in the same position, the pendulum having done quite a bit of moving in the darkness between flashes. If the flashes catch the pendulum at its extreme right position, the pendulum will appear to being pulled, pushed or blown right.

The stroboscopic flashes create visual ambiguity. There are different possible explanations for what the viewer sees. The viewer typically, and often nonconsciously, chooses the explanation that meets his expectations. If you and others saw no movement in a daylight object, it would be considered bizarre for you to proclaim that the object was swinging back and forth. However, this bizarre proclamation would be correct with the apparently motionless pendulum.

Do these three snapshots show a moving or still pendulum? It's impossible to tell.

* * * *

All movies as stroboscopic-like illusions

Despite audience perception, movies don't show continuous, real movement of a deer running, a car racing or people conversing, but a series of snapshots of the movement. If you hold up movie film, you will see it is a series of still images lined up side by side, not unlike the panels in a newspaper comic strip. When the film is shot and shown at the proper speed, the viewer's mind incorrectly interprets the succession of still images as real movement. To the mind, 'realistic movement' seems the most plausible explanation for what it is seeing. This choice is made instantly and nonconsciously and the viewer simply thinks she's watching real, continuous movement.

When the film is too slow, the mind is no longer fooled. The running horse looks choppy and unreal.

* * * *

Ambiguity

Ambiguity is a concept essential to understanding humans, as humans constantly make choices in the face of ambiguous information. Ambiguity means there is more than one possible explanation to something, and the viewer doesn't know, often can't know, which one is correct. In the face of ambiguity, the mind will almost always pick the explanation that meets its expectations and experience. Visual illusions, both moving and still, involve making the wrong pick.

The human mind is designed for speed. Speedy perceptions are essential for living and surviving in the real world, including processing fast movement like a charging lion and rolling bolder. A downside of the speed is there is a fair margin of error. Speed is often synonymous with haste.

* * * *

Ambiguous Movement: The Barber Pole Illusion

There are instances where, due to restricted viewpoint, it is impossible for the viewer to know the direction of movement. A standard example involves the barber pole.

Hung outside the barber shop, a barber pole has diagonal candy cane stripes that are rotated horizontally. However, looking from a particular angle it will appear as if the stripes are moving vertically. Faced with different plausible choices for what it is seeing (possibly moving up, but also possibly moving sideways), the mind takes a pick, one that happens to be wrong.

If you watch a barber pole from different angles, you will alternately perceive the stripes moving vertically and rotating horizontally. Your mind can't make up its mind.

* * * *

As the following three pictures illustrate, even still images can trick the mind into perceiving motion. Their designs match up with the nonconscious brain's template for what is movement.

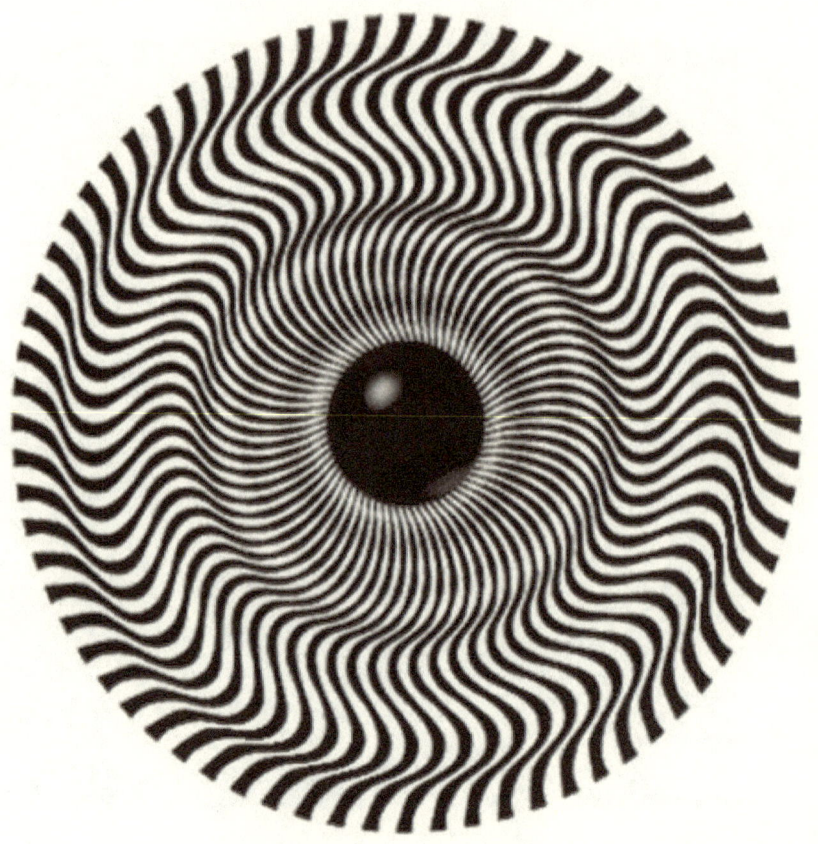

Conceits * David Rudd Cycleback

(c)
Narrative and the Perception
of Still Information

What's the story in this Jan Vermeer painting?

What's the fox doing?

What do you suppose was happening here?

Narrative is an integral part of how humans perceive, identify and judge information, both moving and still, realistic and abstract. A narrative is the conscious and nonconscious story we see and tell about our lives, attach to observed situations and still objects. Narrative includes perception of time, plot, order, causation, mood, action, point of view, emphasis (what is important, what is not), character motives, past and future. When we look at a still photo or painting or a distant couple standing at a street light we perceive a story in progress. We may not know the story, but we take for granted that there is one. A cup on a table isn't just there, there is a history of how it got there, where it will go next. Presumably, a human walked up to the table and placed the cup there, perhaps drank from it. "Who left this dirty cup on the table?!," someone may soon say. "Dirty dishes go in the dishwasher."

We know the earlier fox image was an observer's snapshot of a real living animal in mid movement. A good guess is the fox is/was chasing prey. Did it catch anything? That's a question to ponder.

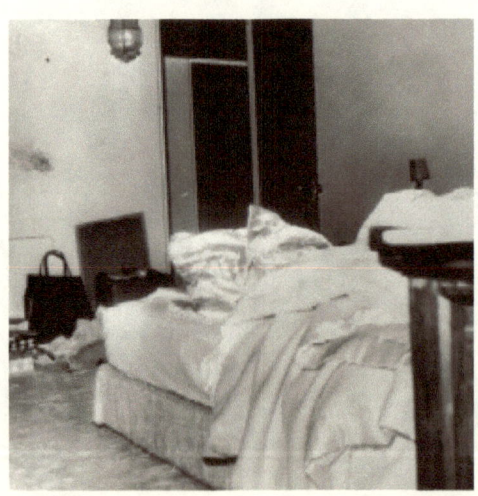

* * * *

If you change the narrative to an image, you change the meaning of the image, at least the perceived meaning. This is why narrative issues are so important. A still image of a man with a knife is generally defined by the narrative-- what he is perceived as going to do with the knife, what he is perceived as having done with the knife. If the narrative is he just cleaned a fish and is taking the knife to the sink, the still image has one meaning. If the narrative is he's looking to hide a murder weapon, the same still has a distinctly different meaning. The accuracy of the narrative is no small issue.

It brings up the question of if a still image can be understood independent of narrative. The two knife narratives were for the same image. Can the cup on the table's identity and meaning be determined as it is? Is how it got there essential to its identity? Humans often like to think they can judge things in a vacuum, without the relativism of past and present and nearby other objects, but is it possible?

Some things are defined by their movement. A cheetah in the wild is defined and identified by how fast it runs. A sidewinder snake is identified because it moves sideways. When it's just laying there, most of us wouldn't know what kind of snake it is.

Much of our narrative is speculative. We can guess but don't know the whole story. The judgment of significance, motives and movements of the players in a scene is influenced by our biases and personal experiences. Different viewers see different stories in the same movie.

Consciously and nonconsciously predicting what will happen is a necessary part of human function. To catch a ball, you don't need to know just where the ball is at any given moment in flight, but correctly anticipate where it will be at later moments.

Narrative is an expression of human's philosophy of time, cause-and-effect, relationships between things. To most humans, nothing is static, but a part of a linear flow. Even still things and still images of things are viewed as part of this flow.

What is particularly interesting is humans apply narratives to abstract images and other information where it is not clear there is a real narrative.

Describe what going on above? Even though this is an abstract combination of dots and lines, most will say this shows two balls racing towards each other. Viewers can even describe what they see as happening before and after this image. However, unlike a movie still or snapshot photo, there is no before or after. As I am the one who created this design, I can assure that this is the only image, the one and only existence of these dots and lines. There is no narrative with this image other than as speculated by the viewer. That it shows balls on a line is itself imagination.

Whether there is a real narrative to the earlier Vermeer painting is debatable. It's not a photographic snapshot of live movement, like with the fox. The narrative and resulting meaning is nothing more than speculation.

As you can see, artistic experience is speculative, theoretical. Art is a symbol and metaphor for something larger and something in the viewers' minds. Art isn't so much interpreted by the viewer as made up. Movement is imagined in the following Matisse, but it doesn't literally exist. Even the artist having imagined movement doesn't make it exist.

A question to consider is is narrative the correct way to judge information? Is it always the correct way? And if it is correct to view information via narrative, is the human narrative the correct narrative? Does all human narrative, even as used by scientists, involve imagination and the associated biases and psychology? Of course, many of these questions we can't answer.

The movement illusions in the previous chapter were all about false narratives. The stroboscopic illusion involves the viewer creating a narrative about movement that differs from reality. The perceiving of the barber pole stripes continually moving up is a false narrative.

The mentally ill often have abnormal narratives. They see and experience the same now and past and future that you and I see, but give the pieces a different causality and relationships, viewpoint, emphasis and soundtrack.

Aleatory Narrative

"Any path is right, if— as according to Bach-- it leads to the divine"— music historian Paul Epstein on J.S. Bach's fugues, to which Bach never gave a playing order.

Aleatory art is art where the finished result is substantially out of the artist's hands. It can involve chance or the musicians' or audience's choice. Many games are aleatory. Monopoly involves the roll of the dice. Poker involves the shuffling of the cards. Aleatocism in art can create fresh, inventive, unexpected results. If the results defies the conventions of plot, narrative and order, that's the point.

J.S. Bach's fugues are aleatory in that he never communicated which order the short musical pieces should be played. They can be played or listened to in any order, take your pick, randomly program the CD player. In the above quote,

Epstein is saying an overall sublime aesthetic result justifies whichever fugue order lead to it. It's reminiscent of the Hindi saying, "Any path that leads to God is correct."

Novelist William S. Burroughs used the so called cut-up aleatory technique. Pages of text were physically cut up and randomly pieced back together, sometimes with text by other authors, creating new and often profoundly surreal meaning and narrative. Burroughs believed this type of collage more closely represented the human experience. Despite the conceit of linearity, humans don't think or experience things linearly, one's thoughts constantly flipping back and forth between past, current and future. Random little events and objects trigger memories and provoke speculation of the future. When you consider buying a can of beans in the grocery isle, you think about past meals and the future meal where these beans might be used. The human ability to identify flowers, shoe brands and people involves comparing the present to memory. Human intelligence and reasoning involves mentally flipping back and forth through time.

Broken Glass is the name of an aleatory computer storytelling technique that intentionally scrambles the tradition linear narrative. It is a computer web page made up of a plethora of small assorted images, often resembling a stained glass window. Each image is linked to a small piece of the story-- a plot, a description, a picture, characterization, whatever. The story's order is determined by the reader blindly clicking on the images.

The facts, scenes, characters, events and days of the week are always constant in Broken Glass, but the aleatory order in which the pieces are read affects the complexion, aesthetics, psychology and meaning. As any great novelist or film director will tell you, how facts are revealed can be as important as the facts themselves. A story told straight foreword is markedly different than the same story told in flashbacks. Knowing what will happen to a character, what she will do and how she will change, effects how you view her in the present. Knowing versus not knowing how the romance will end (or will it end?) effects how the movie goers view the lovers when the first meet, interact. Jumbled up order in and of itself has psychological meaning and symbolism.

Even with a physically bound paper book, the reader chooses the order in which the book is read. Whether or not they realize it, readers are as responsible for the order as the author, though the author usually gets the blame.

William S. Burroughs said the chapters of his novel *Naked Lunch* could be read in any order. That a reader read them 1, 2, 3 had nothing to do with him.

Dictionaries and encyclopedias are aleatory. Excluding the editors and writers, it's possible if not likely that no two people have read the word definitions in a dictionary in the same order.

David Cycleback is an art historian specializing in authentication and fake detection. He is the photography advisor to Becket Media, writer for the Encyclopedia of Nineteenth Century Photography and has advised and examined material for major auction houses. His other books include *Judging the Authenticity of Prints by the Masters, Judging the Authenticity of Photographs* and *Forensic Light: A Beginner's Guide.* His website is cycleback.com

www.ingramcontent.com/pod-product-compliance
Lightning Source LLC
Chambersburg PA
CBHW030927180526
45163CB00002B/490